Lecture Notes in Physics

Volume 881

For further volumes:
www.springer.com/series/5304

The Lecture Notes in Physics

The series Lecture Notes in Physics (LNP), founded in 1969, reports new developments in physics research and teaching—quickly and informally, but with a high quality and the explicit aim to summarize and communicate current knowledge in an accessible way. Books published in this series are conceived as bridging material between advanced graduate textbooks and the forefront of research and to serve three purposes:

- to be a compact and modern up-to-date source of reference on a well-defined topic
- to serve as an accessible introduction to the field to postgraduate students and nonspecialist researchers from related areas
- to be a source of advanced teaching material for specialized seminars, courses and schools

Both monographs and multi-author volumes will be considered for publication. Edited volumes should, however, consist of a very limited number of contributions only. Proceedings will not be considered for LNP.

Volumes published in LNP are disseminated both in print and in electronic formats, the electronic archive being available at springerlink.com. The series content is indexed, abstracted and referenced by many abstracting and information services, bibliographic networks, subscription agencies, library networks, and consortia.

Proposals should be sent to a member of the Editorial Board, or directly to the managing editor at Springer:

Christian Caron
Springer Heidelberg
Physics Editorial Department I
Tiergartenstrasse 17
69121 Heidelberg/Germany
christian.caron@springer.com

Gernot Schaller

Open Quantum Systems
Far from Equilibrium

 Springer

Gernot Schaller
Technical University of Berlin
Berlin, Germany

Additional material to this book can be downloaded from http://extras.springer.com

ISSN 0075-8450   ISSN 1616-6361 (electronic)
Lecture Notes in Physics
ISBN 978-3-319-03876-6   ISBN 978-3-319-03877-3 (eBook)
DOI 10.1007/978-3-319-03877-3
Springer Cham Heidelberg New York Dordrecht London

Library of Congress Control Number: 2013958319

Printed on acid-free paper

Springer is part of Springer Science+Business Media (www.springer.com)

Preface

This book arose from a lecture course on open quantum systems that I had the chance to teach at the Technical University of Berlin. I was asked to give a lecture on my research for an audience that was composed of graduate students specializing in very different areas of physics. Consequently, I had to start with an introduction that generated a common ground. In order to give all students an opportunity to treat hot research topics, I decided not to teach overly sophisticated technical tools. Instead, I tried to make the lecture as self-contained as possible and—with some work involved—straightforward to follow. Presenting that lecture was a fun adventure for me: I had to put my research results into a somewhat wider background and rethink exactly which points were the most important to make. Soon after the actual lecture, I was asked to provide a lecture script for later reference, which triggered the idea for this book.

During the writing of this book, as the research advanced, so did the book; thus, it now contains a few more topics than were treated in the original lecture. However, keeping the original motivation, it aims at providing graduate students or researchers with a little background in quantum theory—what one typically learns during two semesters of quantum theory—with a straight route to the dynamics of open quantum systems. This route is not necessarily easy, since the readers might have to invest some work if they are unfamiliar with certain techniques or topics. Neither can it be claimed to be the only path, and the readers are certainly invited to find and explore possibly simpler or more elegant pathways.

In my opinion, the road to open quantum systems is a very rewarding journey: New decades bring new challenges, and one of the challenges of our decade certainly is to understand and control the behavior of the smallest systems. Just as the steam engine led to the industrial revolution, one can anticipate that nanomachines will not just be useful in existing applications (e.g., drug design and delivery, microfabrication, and DNA construction). Beyond this, they may also yield an unimaginable number of new applications. Nanomachines cannot be described by thermal equilibrium. Therefore, it seems a rewarding enterprise to understand the evolution of open quantum systems when coupled to non-equilibrium reservoirs.

In this book, we will provide several possibilities to treat such non-equilibrium reservoirs. The simplest idea is to compose a non-equilibrium reservoir from sub-systems that are held at different equilibrium states. This approach can only be well motivated in the weak coupling limit. Then, quantum master equations have many favorable properties: These properties enable one to interpret the dynamics of quantum systems coupled to different equilibrium reservoirs similarly to the dynamics of heat engines. Alternatively, we can study strongly coupled quantum systems that—when scaled up in size to the thermodynamic limit with an infinite recurrence time—may assume a non-equilibrium stationary state. Beyond this, there are many more examples of non-equilibrium systems to study. In this book, we will also treat systems subject to external driving and systems that are continuously monitored and controlled, which includes feedback control.

On the technical side, the book provides concepts useful in the presence of the aforementioned situations: multiple reservoirs, non-equilibrium reservoirs, additional monitoring, and feedback control. These methods include master equations, the extraction of full counting statistics from these equations, thermodynamic interpretation of master equations, and of course methods for their solution. It is further demonstrated how the conventional weak coupling limit can be overcome in some cases and how true non-equilibrium reservoirs alter the dynamics. The book contains a number of exercises of varying difficulty, which the reader is invited to solve. The solutions to the exercises are not part of the book, but can be downloaded from the on-line supplement (http://extras.springer.com/ZIP/2014/978-3-319-03877-3.zip). Corrections and suggestions for improvement should be addressed to me:

gernot.schaller@tu-berlin.de

The examples in this contribution have mostly originated from my own research and that of collaborators and students, to whom I would like to express my deepest gratitude. Tobias Brandes, Clive Emary, Massimiliano Esposito, Gerold Kießlich, Thilo Krause, Philipp Strasberg, Christian Nietner, Gabriel Topp, and Malte Vogl have—among many others to whom I apologize for not mentioning them—questioned my views and sharpened my thinking. Without these wonderful people, this book would not have been possible. Any errors are, of course, entirely my own.

Finally, I would like to apologize to my wife and my little daughters for being a distracted husband and father during the writing of this book. After all, it is the joy you bring that keeps me going.

Berlin, Germany   Gernot Schaller

Contents

Chapter 1
Dynamics of Open Quantum Systems

Abstract This chapter provides a brief introduction to quantum systems that are coupled to large reservoirs. It aims to remind the reader of well-known concepts necessary for the understanding of the book and does not claim to provide a self-contained introduction. It starts with a brief summary of the conventions used in the book and then introduces master equations with some examples. This also requires us to introduce the density matrix: among other things, we discuss its evolution in a closed system and under measurements. To connect to system-reservoir theories, we also review the definition of the tensor product and the partial trace. Finally, we introduce the Lindblad form of a quantum master equation and discuss its properties before closing with some remarks on the superoperator representation of master equations.

With the tremendous advances during the last century in our ability to prepare and control the smallest systems, quantum theory has proven extremely successful. This evolution has not only been driven by mere interest in basic principles. Perfect control of quantum systems would also allow one to build extremely powerful computers that could solve special problems such as number factoring [1], database search [2], or simply simulation of other quantum systems [3, 4] much faster than we can do with classical computers. Unfortunately, the promises of quantum computation have turned out to be hard to keep, since the fragile quantum coherence necessary for quantum computation to work usually rapidly decays. This process—commonly termed decoherence [5–7]—is induced by the presence of reservoirs that can significantly alter the true quantum dynamics. With the sophisticated experimental setups in present-day proof-of-principle implementations of quantum computers [8] or quantum simulators [9–11], these reservoirs usually cannot be assumed to be in thermal equilibrium.

The coupling between a quantum system and a structured non-equilibrium environment can however also be seen as a chance: the smallest quantum systems can also be seen as nanomachines that exchange energy and matter with their surroundings. From a thermodynamic viewpoint, such nanomachines are coupled to an environment that is out of equilibrium and might thus be able to perform useful tasks such as generating electrical current from a heat gradient [12]. Alternatively, they could function as heating or cooling devices [13].

G. Schaller, *Open Quantum Systems Far from Equilibrium*, Lecture Notes in Physics 881,   1
DOI 10.1007/978-3-319-03877-3_1,
© Springer International Publishing Switzerland 2014

In either case, the effect of non-equilibrium environments on a quantum system is a topic that deserves to be thoroughly understood. This book provides some basic steps towards a description of open quantum systems subject to non-equilibrium environments.

1.1 Conventions

Altogether, we will use the following conventions without further notice in the book.

Planck's constant $\hbar = 1.0546 \times 10^{-34}$ Js will be set to one; i.e., we will absorb it in the Hamiltonian of every considered system. This implies that all energies will have dimensions of inverse time. Similarly, Boltzmann's constant $k_B = 1.3806 \times 10^{-23}$ J/K will also not occur in this book; it will be hidden in the inverse temperature $\beta = 1/(k_B T)$ with temperature T.

The quantity $[A, B] \equiv AB - BA$ denotes the commutator between two operators A and B, whereas $\{A, B\} = AB + BA$ denotes their anti-commutator.

Operators in the interaction picture will be written by boldface symbols $\boldsymbol{O}(t) = e^{+iH_0 t} O e^{-iH_0 t}$, with H_0 and t denoting the free Hamiltonian and time, respectively.

We will represent superoperators, i.e., linear operations on operators, by calligraphic symbols. For example, the linear operation $\mathscr{K}[O] = \sum_{ij} K_i O K_j$ on the operator O will—after short notice—be denoted by $\mathscr{K} O$.

Throughout the book, we will denote the Fermi–Dirac distribution (or just the Fermi function) of a particular reservoir α by

$$f_\alpha(\omega) = \frac{1}{e^{\beta_\alpha(\omega - \mu_\alpha)} + 1}, \tag{1.1}$$

where β_α and μ_α represent the inverse temperature and chemical potential of the reservoir α, respectively. Similarly, we will denote the Bose–Einstein distribution of bosonic reservoirs by

$$n_\alpha(\omega) = \frac{1}{e^{\beta_\alpha(\omega - \mu_\alpha)} - 1}, \tag{1.2}$$

where we will however mostly consider $\mu_\alpha = 0$.

Finally, we mention that only a few abbreviations will be used in the book. The ones to remember are single electron transistor (SET), double quantum dot (DQD), quantum point contact (QPC), Kubo–Martin–Schwinger (KMS), and Baker–Campbell–Hausdorff (BCH).

1.2 Evolution of Closed Systems

Before we start with the non-equilibrium, we will briefly review closed quantum systems. The dynamics of such a closed quantum system can already be complicated enough, since the evolution of its state vector $|\Psi\rangle$ obeys the Schrödinger equation

$$|\dot{\Psi}(t)\rangle = -iH(t)|\Psi(t)\rangle, \tag{1.3}$$

where we have absorbed the Planck constant $\hbar$ in the Hamiltonian $H(t)$. A time-dependent Hamiltonian in the Schrödinger picture would mean that the system is actually not really closed: changing the parameters of the Hamiltonian normally requires an interaction with the outside world. However, time-dependent Hamiltonians may also arise in transformed pictures, e.g., when a time-dependent unitary transformation $|\tilde{\Psi}(t)\rangle = e^{+iH_0t}|\tilde{\Psi}(t)\rangle$ is applied to Eq. (1.3) with an initially time-independent Hamiltonian.

Exercise 1.1 (Transformation to the interaction picture) Assuming a time-independent Hamiltonian $H = H_0 + V$, show that the Schrödinger equation in the interaction picture becomes

$$|\dot{\tilde{\Psi}}(t)\rangle = -iV(t)|\tilde{\Psi}(t)\rangle, \tag{1.4}$$

where $V(t) = e^{+iH_0t}Ve^{-iH_0t}$ denotes the time-dependent Hamiltonian and $|\tilde{\Psi}(t)\rangle = e^{+iH_0t}|\Psi(t)\rangle$ the state vector in the interaction picture.

The Schrödinger equation is formally solved by the unitary propagator

$$U(t) = \hat{\tau}\exp\left\{-i\int_0^t H(t')\,dt'\right\}, \tag{1.5}$$

with the time-ordering operator $\hat{\tau}$. Time ordering sorts time-dependent operators depending on their time argument; i.e., formally it acts as

$$\hat{\tau}\mathbf{O}(t_1)\mathbf{O}(t_2) = \Theta(t_1 - t_2)\mathbf{O}(t_1)\mathbf{O}(t_2) + \Theta(t_2 - t_1)\mathbf{O}(t_2)\mathbf{O}(t_1) \tag{1.6}$$

with the Heaviside theta function

$$\Theta(x) = \begin{cases} 1: & x > 0, \\ 1/2: & x = 0, \\ 0: & x < 0. \end{cases} \tag{1.7}$$

Its role in the time evolution operator can however also be defined by the time derivative

$$\dot{U}(t) = -iH(t)U(t). \tag{1.8}$$

In the case of a time-independent Hamiltonian however, time ordering is not necessary and we simply obtain $U(t) = e^{-iHt}$. This neglect of time ordering is possible only when the commutator of the Hamiltonian with itself vanishes at different times $[H(t), H(t')] = 0$. In the general case however, the study of time-dependent Hamiltonians is usually quite difficult and is normally restricted to periodic [14] or adiabatic [15] time dependencies. Turning the question around, it is simpler to take a time-dependent trajectory of the state vector and to obtain a corresponding time-dependent Hamiltonian [16]. Unfortunately, this is often not the question asked in the experimental setup.

In any case however, unitary evolution $(U^\dagger(t)U(t) = \mathbf{1})$ means that the information about the initial state is conserved in every solution to the Schrödinger equation. A unitarily evolving system cannot evolve towards a single stationary state, since from that state the information about the initial configuration cannot be extracted. For a constant Hamiltonian, we may expand the initial state in the eigenstates $H|n\rangle = E_n|n\rangle$ of the Hamiltonian, and the time-dependent solution to the Schrödinger equation is then simply given by $|\Psi(t)\rangle = \sum_n c_n^0 \exp\{-iE_n t\}|n\rangle$. For a finite number of system energies E_n, this will always evolve periodically and thus return to its initial state after some recurrence time. When the system becomes large however, approximate notions of a stationary state in a closed quantum system exist [17].

Furthermore, realistic quantum systems can usually not be regarded as closed; i.e., they are not perfectly isolated from their environment (composed of thermal reservoirs, detectors, and other things). The naive approach of simply simulating the evolution of both the system and its environment is unfortunately prohibitive. With increasing size, the complexity to simulate a quantum system grows exponentially, and a typical reservoir with $\mathcal{O}\{10^{23}\}$ degrees of freedom would in the simplest case require the storage of $\mathcal{O}\{2^{10^{23}}\}$ bits, which is completely impossible. With our limited abilities one should therefore be content with a theory that describes only a small part of our universe—conventionally called the system. In this restricted subspace, the dynamics may no longer be expected to be unitary. That is, a simple time dependence of external parameters in the Hamiltonian cannot account for the observed dynamics, which the Schrödinger equation (1.3) will fail to predict. In such cases, the system can no longer be described by a pure state $|\Psi\rangle$, and the density matrix formalism is required. This formalism will be introduced in the following sections.

1.3 Master Equations

1.3.1 Definition

Many processes in nature are stochastic. In classical physics, this may be due to our incomplete knowledge of the system. Due to the unknown microstate of, e.g., a gas in a box, the collisions of gas particles with the domain wall will appear random. In quantum theory, the Schrödinger equation (1.3) itself involves amplitudes rather than observables in the lowest level, and measurement of observables will yield a stochastic outcome. In order to understand such processes in great detail, such random events must be included via a probabilistic description. For dynamical systems, probabilities associated with measurement outcomes may evolve in time, and the determining equation for such a process is called a master equation.

Definition 1.1 (Master equation) A master equation is a first-order differential equation describing the time evolution of probabilities, e.g., for discrete events $k \in \{1, \ldots, N\}$

$$\frac{dP_k}{dt} = \sum_\ell [T_{k\ell} P_\ell - T_{\ell k} P_k], \tag{1.9}$$

where the $T_{k\ell} \geq 0$ are transition rates from state (measurement event) ℓ to state (measurement event) k. Since it is completely defined by the transition rates, it is also termed the rate equation.

The master equation is said to satisfy detailed balance when for the stationary state $\bar{P}_i$ the equality $T_{k\ell} \bar{P}_\ell = T_{\ell k} \bar{P}_k$ holds for all terms separately.

The transition rates must be positive and may in principle also depend on time. When the transition matrix $T_{k\ell}$ is symmetric, all processes are reversible at the level of the master equation description.

Often, master equations are phenomenologically motivated and not derived from first principles. However, in most examples discussed in this book we will use master equations that can be derived from a microscopic underlying model. We will see later that, in its standard form, the Markovian quantum master equation may not only involve probabilities (diagonals of the density matrix, termed populations) but also further auxiliary values (off-diagonal entries, termed coherences). However, it is possible to transform such master equations to a rate equation representation in a suitable basis. Therefore, we will use the term master equation in this book in a somewhat wider sense as an equation that provides the time equation of probabilities.

It is straightforward to show that the master equation conserves the total probability

$$\sum_k \frac{dP_k}{dt} = \sum_{k\ell} (T_{k\ell} P_\ell - T_{\ell k} P_k) = \sum_{k\ell} (T_{\ell k} P_k - T_{\ell k} P_k) = 0. \tag{1.10}$$

Beyond this, all probabilities must remain positive, which is also respected by a rate equation with positive rates: evidently, the solution of the master equation is continuous, such that when initialized with valid probabilities $0 \leq P_i(0) \leq 1$ all probabilities are non-negative initially. Let P_k be the first probability that approaches zero at some time t (when all other probabilities are non-negative). Its time derivative is then given by

$$\left. \frac{dP_k}{dt} \right|_{P_k=0} = +\sum_\ell T_{k\ell} P_\ell \geq 0, \tag{1.11}$$

which simply implies that $P_k(t)$ will increase. In effect, any probability will be repelled from zero, such that negative probabilities are impossible with positive rates.

Finally, the probabilities must remain smaller than one throughout the evolution. This however follows immediately from $\sum_k P_k = 1$ and $P_k \geq 0$ by contradiction.

In conclusion, a master equation of the form (1.9) automatically preserves the sum of probabilities and also keeps $0 \leq P_i(t) \leq 1$—with a valid initialization provided. That is, under the evolution of a rate equation, probabilities remain probabilities.

1.3.2 Examples

1.3.2.1 Fluctuating Two-Level System

Let us consider a system of two possible states, to which we associate the time-dependent probabilities $P_0(t)$ and $P_1(t)$. These events could for example be the two conformations of a molecule, the configurations of a spin, the ground and excited states of an atom, etc. To introduce some dynamics, let the transition rate from $0 \rightarrow 1$ be denoted by $T_{10} > 0$ and the inverse transition rate $1 \rightarrow 0$ be denoted by $T_{01} > 0$. This implies that the conditional probability to end up in the state 1 at time $(t + \Delta t)$ provided that at time t one is in the state 0, is for sufficiently small time intervals Δt given by $T_{10}\Delta t$. The associated master equation is then a first-order differential equation given by

$$\frac{d}{dt}\begin{pmatrix} P_0 \\ P_1 \end{pmatrix} = \begin{pmatrix} -T_{10} & +T_{01} \\ +T_{10} & -T_{01} \end{pmatrix}\begin{pmatrix} P_0 \\ P_1 \end{pmatrix}. \tag{1.12}$$

We note that in the matrix representation, conservation of the trace is fulfilled when the entries in all columns of the rate matrix add up to zero. This can easily be shown to hold more generally.

Exercise 1.2 (Temporal dynamics of a two-level system) Calculate the solution of Eq. (1.12). What is the stationary state? Show that detailed balance is satisfied.

1.3.2.2 Diffusion Equation

Consider an infinite chain of coupled compartments. Now suppose that, along the chain, molecules may move from one compartment to another with a transition rate $T > 0$ that is unbiased, i.e., symmetric in all directions as depicted in Fig. 1.1. The evolution of probabilities obeys the infinite-size master equation

$$\dot{P}_i(t) = T P_{i-1}(t) + T P_{i+1}(t) - 2T P_i(t)$$

$$= T \Delta x^2 \frac{P_{i-1}(t) + P_{i+1}(t) - 2P_i(t)}{\Delta x^2}, \tag{1.13}$$

which converges as $\Delta x \rightarrow 0$ and $T \rightarrow \infty$ such that $D = T \Delta x^2$ remains constant to the partial differential equation

$$\frac{\partial P(x,t)}{\partial t} = D \frac{\partial^2 P(x,t)}{\partial x^2} \quad \text{with } D = T \Delta x^2, \tag{1.14}$$

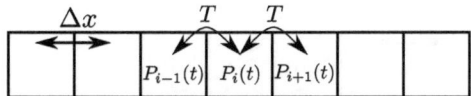

Fig. 1.1 Sketch of a chain of compartments, between which a transition is possible with isotropic and uniform rate $T > 0$. In the limit when the compartment size $\Delta x \to 0$ and $T \to \infty$ such that $\Delta x^2 T = D$ remains constant, the dynamics of the probabilities is described by the diffusion equation

where D is the diffusion constant. Such diffusion equations are used to describe the distribution of chemicals in a solution in the highly diluted limit, the kinetic dynamics of bacteria, and further undirected transport processes. From our analysis of master equations, we can immediately conclude that the diffusion equation preserves positivity and total norm, i.e., $P(x,t) \geq 0$ and $\int_{-\infty}^{+\infty} P(x,t)\,dx = 1$. Note that it is straightforward to generalize to the higher dimensional case.

One can now think of microscopic models where the hopping rates in different directions are not equal (drift) and may also depend on the position (a spatially dependent diffusion coefficient). A position-dependent hopping rate may, e.g., result from a heterogeneous medium through which transport occurs, whereas a difference in the directionality may result from an applied external potential (e.g., in the case of electrons) or some intrinsic preference of the considered species (e.g., in the case of chemotactically active bacteria sensing a present chemical gradient). A corresponding model (in a next-neighbor approximation) would be given by

$$\dot{P}_i = T_{i,i-1} P_{i-1}(t) + T_{i,i+1} P_{i+1}(t) - (T_{i-1,i} + T_{i+1,i}) P_i(t), \qquad (1.15)$$

where $T_{a,b}$ denotes the rate of jumping from b to a; see also Fig. 1.2. An educated guess is given by the ansatz

$$
\begin{aligned}
\frac{\partial P}{\partial t} &= \frac{\partial^2}{\partial x^2}\big[A(x)P(x,t)\big] + \frac{\partial}{\partial x}\big[B(x)P(x,t)\big] \\
&\equiv \frac{A_{i-1}P_{i-1} - 2A_i P_i + A_{i+1}P_{i+1}}{\Delta x^2} + \frac{B_{i+1}P_{i+1} - B_{i-1}P_{i-1}}{2\Delta x} \\
&= \left[\frac{A_{i-1}}{\Delta x^2} - \frac{B_{i-1}}{2\Delta x}\right]P_{i-1} - \frac{2A_i}{\Delta x^2}P_i + \left[\frac{A_{i+1}}{\Delta x^2} + \frac{B_{i+1}}{2\Delta x}\right]P_{i+1}, \qquad (1.16)
\end{aligned}
$$

which is equivalent to our master equation when

$$A_i = \frac{\Delta x^2}{2}[T_{i-1,i} + T_{i+1,i}], \qquad B_i = \Delta x[T_{i-1,i} - T_{i+1,i}]. \qquad (1.17)$$

We conclude that the Fokker–Planck equation

$$\frac{\partial P}{\partial t} = \frac{\partial^2}{\partial x^2}\big[A(x)P(x,t)\big] + \frac{\partial}{\partial x}\big[B(x)P(x,t)\big] \qquad (1.18)$$

with $A(x) \geq 0$ preserves norm and positivity of the probability distribution $P(x,t)$.

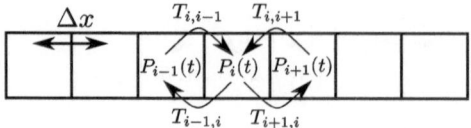

Fig. 1.2 Sketch of a chain of compartments, between which a transition is possible with differing rates $T_{ij} \geq 0$. In contrast to Fig. 1.1, the hopping rates are not uniform, $T_{ij} \neq T$, and may be anisotropic, $T_{ij} \neq T_{ji}$

Exercise 1.3 (Reaction-diffusion equation) Along a linear chain of compartments, consider the master equation for two species

$$\dot{P}_i = T\big[P_{i-1}(t) + P_{i+1}(t) - 2P_i(t)\big] - \gamma P_i(t),$$
$$\dot{p}_i = \tau\big[p_{i-1}(t) + p_{i+1}(t) - 2p_i(t)\big] + \gamma P_i(t),$$

where $P_i(t)$ may denote the concentration of a molecule that irreversibly reacts with chemicals in the solution to an inert form characterized by $p_i(t)$. To which partial differential equation does the master equation map?

In some cases, the probabilities may not only depend on the probabilities themselves, but also on external parameters, which appear then in the master equation. Here, we will use the term master equation for any equation describing the time evolution of probabilities; i.e., auxiliary variables may appear in the master equation.

1.3.2.3 Cell Culture Growth

Consider a population of identical cells, where each cell may divide (proliferate) with a rate α. These cells live in a constrained geometry (e.g., a Petri dish) that admits at most K cells due to some limitations (space, nutrient supply, etc.). Let $P_i(t)$ denote the probability of having i cells in the Petri dish. Assuming that the proliferation rate α is sufficiently small, we can easily set up a master equation:

$$\dot{P}_0 = 0,$$
$$\dot{P}_1 = -1 \cdot \alpha \cdot P_1,$$
$$\dot{P}_2 = -2 \cdot \alpha \cdot P_2 + 1 \cdot \alpha \cdot P_1,$$
$$\vdots$$

$$\dot{P}_\ell = -\ell \cdot \alpha \cdot P_\ell + (\ell - 1) \cdot \alpha \cdot P_{\ell-1},$$

$$\vdots$$

(1.19)

$$\dot{P}_{K-1} = -(K-1) \cdot \alpha \cdot P_{K-1} + (K-2) \cdot \alpha \cdot P_{K-2},$$

$$\dot{P}_K = +(K-1) \cdot \alpha P_{K-1}.$$

The prefactors in front of the bare rates arise since any of the ℓ cells may proliferate. Arranging the probabilities in a single vector, this may also be written as $\dot{\boldsymbol{P}} = \mathscr{L}\boldsymbol{P}$, where the band-diagonal matrix $\mathscr{L}$ contains the rates. When we have a single cell as the initial condition (full knowledge), i.e., $P_1(0) = 1$ and $P_{\ell \neq 1}(0) = 0$, one can change the carrying capacity $K = \{1, 2, 3, 4, \ldots\}$ and solve for each K the resulting system of differential equations for the expectation value of $\langle \ell \rangle = \sum_{\ell=1}^{K} \ell P_\ell(t)$. These solutions may then be generalized to

$$\langle \ell \rangle = e^{+\alpha t}\left[1 - \left(1 - e^{-\alpha t}\right)^K\right]. \tag{1.20}$$

Similarly, one can compute the expectation value of $\langle \ell^2 \rangle$.

Exercise 1.4 (Cell culture growth) Confirm the validity of Eq. (1.20).

This result can be compared with the logistic growth equation, obtained from the solution of the differential equation

$$\dot{N} = \alpha\left(1 - \frac{N}{K}\right)N, \tag{1.21}$$

which means that initially cell growth is just given by the bare proliferation rate α and then smoothly reduced when the population approaches the carrying capacity K.

Exercise 1.5 (Logistic growth equation) Solve Eq. (1.21).

However, one may not only be interested in the evolution by mean values. Sometimes, rare events become quite important (e.g., a benign tumor cell turning malignant), in particular when they are strengthened in the following dynamics. Then it is also useful to obtain some information about the spread of single trajectories from the mean. In the case of a rate equation only involving the probabilities, as in Eq. (1.19), it is possible to also generate single trajectories from the master equation by using Monte Carlo simulation. Suppose that, at time t, the system is in the state ℓ, i.e., $P_\alpha(t) = \delta_{\ell\alpha}$. After a sufficiently short time Δt, the probabilities of being in a different state read as

$$\boldsymbol{P}(t + \Delta t) \approx [\mathbf{1} + \Delta t \mathscr{L}]\boldsymbol{P}(t) + \mathscr{O}\{\Delta t\}^2, \tag{1.22}$$

which for our simple example boils down to

$$P_\ell(t + \Delta t) \approx (1 - \ell\alpha\Delta t)P_\ell(t), \qquad P_{\ell+1}(t + \Delta t) \approx +\ell\alpha\Delta t\, P_\ell(t). \tag{1.23}$$

To simulate a single trajectory, one may now simply draw a random number $\sigma \in [0, 1]$: the probability that a cell divides during this small time interval is given

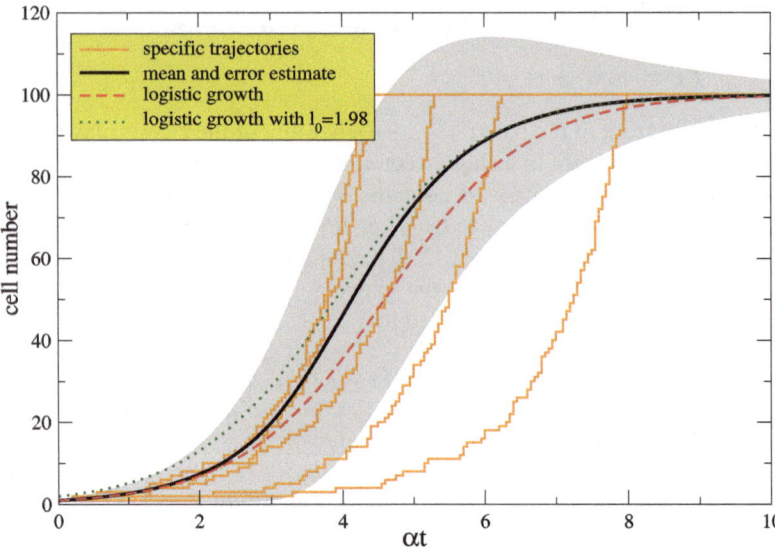

Fig. 1.3 Population dynamics for the linear master equation (*black curve* $\langle \ell \rangle$ and *shaded area* determined from $\sqrt{\langle \ell^2 \rangle - \langle \ell \rangle^2}$) and for the logistic growth equation (*dashed red curve*) for carrying capacity $K = 100$. For identical initial and final states, the master equation solution overshoots the logistic growth curve. A slight modification (*dotted green curve*) of the initial condition in the logistic growth curve yields the same long-term asymptotics. An average of many specific trajectories would converge towards the *black curve*

by $P_{\text{jump}} = \ell \alpha \Delta t \ll 1$. If the random number $\sigma \leq P_{\text{jump}}$, we assume that the transition $P_\ell \rightarrow P_{\ell+1}$ has occurred, and we may set $P_\alpha(t + \Delta t) = \delta_{\ell+1,\alpha}$. In contrast, when $\alpha > P_{\text{jump}}$, we assume that no transition has occurred and therefore remain at $P_\alpha(t + \Delta t) = \delta_{\ell\alpha}$. In any case, the simulation keeps track of the actual state of the system as if the cell number were regularly measured at intervals Δt. The ensemble average of many such trajectories will yield the mean evolution predicted by the master equation; see Fig. 1.3. The figure demonstrates that single trajectories may look quite different from the solution of the master equation. Furthermore, the mean and standard deviation (shaded area) may hide important information about single trajectories: in this case, single trajectories must always be bounded by the carrying capacity K. The ensemble averages of trajectories must however coincide with the rate equation solution.

1.4 Density Matrix Formalism

1.4.1 Density Matrix

Suppose one wants to describe a quantum system, where the system state is not exactly known. That is, there is an ensemble of known states $\{|\Phi_i\rangle\}$, but there is uncertainty regarding in which of these states the system resides. Such systems can be conveniently described by a density matrix.

Definition 1.2 (Density matrix) Any density matrix can be written as

$$\rho = \sum_i p_i |\Phi_i\rangle\langle\Phi_i|, \tag{1.24}$$

where $0 \le p_i \le 1$ denote the probabilities of being in the state $|\Phi_i\rangle$ with $\sum_i p_i = 1$. In general, the states are not required to be orthogonal, i.e., $\langle\Phi_i|\Phi_j\rangle \neq \delta_{ij}$. Formally, any matrix fulfilling the properties

- self-adjointness: $\rho^\dagger = \rho$
- normalization: $\mathrm{Tr}\{\rho\} = 1$
- positivity: $\langle\Psi|\rho|\Psi\rangle \ge 0$ for all vectors Ψ

can be interpreted as a valid density matrix.

For a pure state, one has $p_{\bar{i}} = 1$ and thereby $\rho = |\Phi_{\bar{i}}\rangle\langle\Phi_{\bar{i}}|$ for some particular $\bar{i}$.

Thus, a density matrix is pure if and only if $\rho = \rho^2$.

The expectation value of an operator for a known state $|\Psi\rangle$

$$\langle A\rangle = \langle\Psi|A|\Psi\rangle \tag{1.25}$$

can be obtained conveniently from the corresponding pure density matrix $\rho = |\Psi\rangle\langle\Psi|$ by simply computing the trace (sum of diagonal elements) of $A\rho$:

$$\langle A\rangle \equiv \mathrm{Tr}\{A\rho\} = \mathrm{Tr}\{\rho A\} = \mathrm{Tr}\{A|\Psi\rangle\langle\Psi|\}$$

$$= \sum_n \langle n|A|\Psi\rangle\langle\Psi|n\rangle = \langle\Psi|\left(\sum_n |n\rangle\langle n|\right)A|\Psi\rangle$$

$$= \langle\Psi|A|\Psi\rangle. \tag{1.26}$$

In the first line above, we simply stated an important property of the trace: its invariance under cyclic permutations of its arguments. When the state is not exactly known, but its probability distribution is, the expectation value is obtained by computing the weighted average

$$\langle A\rangle = \sum_i P_i \langle\Phi_i|A|\Phi_i\rangle, \tag{1.27}$$

where P_i denotes the probability of being in state $|\Phi_i\rangle$. The definition of obtaining expectation values by calculating traces of operators with the density matrix is also consistent with mixed states

$$\langle A \rangle \equiv \mathrm{Tr}\{A\rho\} = \mathrm{Tr}\left\{ A \sum_i p_i |\Phi_i\rangle\langle\Phi_i| \right\} = \sum_i p_i \, \mathrm{Tr}\{A|\Phi_i\rangle\langle\Phi_i|\}$$

$$= \sum_i p_i \sum_n \langle n|A|\Phi_i\rangle\langle\Phi_i|n\rangle = \sum_i p_i \langle\Phi_i| \left(\sum_n |n\rangle\langle n| \right) A|\Phi_i\rangle$$

$$= \sum_i p_i \langle\Phi_i|A|\Phi_i\rangle. \tag{1.28}$$

Exercise 1.6 (Superposition versus localized states) Calculate the density matrix for a statistical mixture in the states $|0\rangle$ and $|1\rangle$ with probability $p_0 = 3/4$ and $p_1 = 1/4$. What is the density matrix for a statistical mixture of the superposition states $|\Psi_a\rangle = \sqrt{3/4}|0\rangle + \sqrt{1/4}|1\rangle$ and $|\Psi_b\rangle = \sqrt{3/4}|0\rangle - \sqrt{1/4}|1\rangle$ with probabilities $p_a = p_b = 1/2$?

1.4.2 Dynamical Evolution of a Density Matrix

1.4.2.1 Continuous Evolution

The evolution of a pure state vector in a closed quantum system is described by the evolution operator $U(t)$, as, e.g., for the Schrödinger equation (1.3) the time evolution operator (1.5) may be defined as the solution to the operator equation $\dot{U}(t) = -\mathrm{i}H(t)U(t)$. For constant $H(t) = H$, we simply have the solution $U(t) = e^{-\mathrm{i}Ht}$. Similarly, a pure state density matrix $\rho = |\Psi\rangle\langle\Psi|$ would evolve according to the von Neumann equation

$$\dot{\rho} = -\mathrm{i}\big[H(t), \rho(t) \big] \tag{1.29}$$

with the formal solution $\rho(t) = U(t)\rho(0)U^\dagger(t)$; compare Eq. (1.5). This simply means that for pure states, the von Neumann equation yields the same dynamics as the Schrödinger equation, and pure states remain pure under unitary evolution.

When we apply the very same evolution equation (1.29) to a density matrix that is not pure, we obtain

$$\rho(t) = \sum_i p_i U(t)|\Phi_i\rangle\langle\Phi_i|U^\dagger(t). \tag{1.30}$$

This equation implies that transitions between the (now time-dependent) state vectors $|\Phi_i(t)\rangle = U(t)|\Phi_i\rangle$ are impossible with unitary evolution. These are exactly the state vectors that one would have obtained from the Schrödinger equation by initializing with the initial state $|\Phi_i\rangle$. Therefore, the von Neumann evolution equation yields the same dynamics as an ensemble average of the Schrödinger equation solutions corresponding to the different initial states.

Exercise 1.7 (Preservation of density matrix properties by unitary evolution) Show that the von Neumann equation (1.29) preserves self-adjointness, trace, and positivity of the density matrix.

1.4.2.2 Measurement

The measurement process can also be generalized similarly. For a quantum state $|\Psi\rangle$, measurements are described by a set of measurement operators $\{M_m\}$, each corresponding to a certain measurement outcome, and with the completeness relation $\sum_m M_m^\dagger M_m = 1$. The probability of obtaining result m is given by

$$P_m = \langle \Psi | M_m^\dagger M_m | \Psi \rangle \tag{1.31}$$

and after the measurement with outcome m, the quantum state is collapsed:

$$|\Psi\rangle \xrightarrow{m} \frac{M_m |\Psi\rangle}{\sqrt{\langle \Psi | M_m^\dagger M_m | \Psi \rangle}}. \tag{1.32}$$

The projective measurement is just a special case of that with $M_m = |m\rangle\langle m|$.

Definition 1.3 (Measurements with density matrix) For a set of measurement operators $\{M_m\}$ corresponding to different outcomes m and obeying the completeness relation $\sum_m M_m^\dagger M_m = 1$, the probability of obtaining result m is given by

$$P_m = \mathrm{Tr}\{M_m^\dagger M_m \rho\}, \tag{1.33}$$

and the action of measurement on the density matrix—provided that result m was obtained—can be summarized as

$$\rho \xrightarrow{m} \rho' = \frac{M_m \rho M_m^\dagger}{\mathrm{Tr}\{M_m^\dagger M_m \rho\}}. \tag{1.34}$$

The set of measurement operators is also the called positive operator-valued measure (POVM).

It is therefore straightforward to see that the descriptions using the Schrödinger equation and the von Neumann equation with the respective measurement postulates are equivalent. The density matrix formalism conveniently includes statistical mixtures in the description, since it automatically performs the averaging over different initial conditions. Unfortunately, this comes at the cost of quadratically increasing the number of state variables.

Exercise 1.8 (Preservation of density matrix properties by measurement) Show that the measurement postulate preserves self-adjointness, trace, and positivity of the density matrix.

1.4.2.3 Most General Evolution

Finally, we mention here that the most general evolution preserving all the nice properties of a density matrix is called a Kraus map [18]. A density matrix ρ (hermitian, positive definite, and with trace one) can be mapped to another density matrix ρ' via

$$\rho' = \sum_{\alpha\beta} \gamma_{\alpha\beta} A_\alpha \rho A_\beta^\dagger, \quad \text{with } \sum_{\alpha\beta} \gamma_{\alpha\beta} A_\beta^\dagger A_\alpha = \mathbf{1}, \tag{1.35}$$

where the prefactors $\gamma_{\alpha\beta}$ form a hermitian ($\gamma_{\alpha\beta} = \gamma_{\beta\alpha}^*$) and positive definite ($\sum_{\alpha\beta} x_\alpha^* \gamma_{\alpha\beta} x_\beta \geq 0$ or equivalently all eigenvalues of $(\gamma_{\alpha\beta})$ are non-negative) matrix. It is straightforward to see that the above map preserves trace and hermiticity of the density matrix. In addition, ρ' also inherits the positivity from $\rho = \sum_n P_n |n\rangle\langle n|$

$$\langle \Psi | \rho' | \Psi \rangle = \sum_{\alpha\beta} \gamma_{\alpha\beta} \langle \Psi | A_\alpha \rho A_\beta^\dagger | \Psi \rangle = \sum_n P_n \sum_{\alpha\beta} \gamma_{\alpha\beta} \langle \Psi | A_\alpha | n \rangle \langle n | A_\beta^\dagger | \Psi \rangle$$

$$= \sum_n \underbrace{P_n}_{\geq 0} \underbrace{\sum_{\alpha\beta} (\langle n | A_\alpha^\dagger | \Psi \rangle)^* \gamma_{\alpha\beta} \langle n | A_\beta^\dagger | \Psi \rangle}_{\geq 0} \geq 0. \tag{1.36}$$

Since the matrix $\gamma_{\alpha\beta}$ is hermitian, it can be diagonalized by a suitable unitary transformation, and we introduce the new operators $A_\alpha = \sum_{\alpha'} U_{\alpha\alpha'} \bar{K}_{\alpha'}$:

$$\rho' = \sum_{\alpha\beta} \sum_{\alpha'\beta'} \gamma_{\alpha\beta} U_{\alpha\alpha'} \bar{K}_{\alpha'} \rho U_{\beta\beta'}^* \bar{K}_{\beta'}^\dagger = \sum_{\alpha'\beta'} \bar{K}_{\alpha'} \rho \bar{K}_{\beta'}^\dagger \underbrace{\sum_{\alpha\beta} U_{\alpha\alpha'} \gamma_{\alpha\beta} U_{\beta\beta'}^*}_{\gamma_{\alpha'} \delta_{\alpha'\beta'}}$$

$$= \sum_\alpha \gamma_\alpha \bar{K}_\alpha \rho \bar{K}_\alpha^\dagger, \tag{1.37}$$

where $\gamma_\alpha \geq 0$ represent the eigenvalues of the matrix $(\gamma_{\alpha\beta})$. Since these are by construction positive, we introduce further new operators $K_\alpha = \sqrt{\gamma_\alpha} \bar{K}_\alpha$ to obtain the simplest representation of a Kraus map.

Definition 1.4 (Kraus map) The map

$$\rho(t + \Delta t) = \sum_\alpha K_\alpha(t, \Delta t) \rho(t) K_\alpha^\dagger(t, \Delta t) \tag{1.38}$$

with Kraus operators $K_\alpha(t, \Delta t)$ obeying the relation $\sum_\alpha K_\alpha^\dagger(t, \Delta t) K_\alpha(t, \Delta t) = \mathbf{1}$ preserves hermiticity, trace, and positivity of the density matrix.

Obviously, both unitary evolution and evolution under measurement are just special cases of a Kraus map. Though Kraus maps are heavily used in quantum information [5], they are not often very easy to interpret. For example, it is not straightforward to identify the unitary and the nonunitary part induced by the Kraus map.

1.4.3 Tensor Product

The greatest advantage of the density matrix formalism is visible when quantum systems composed of several subsystems are considered. Then, a tensor product is required to construct the Hilbert space of the combined system. Roughly speaking, it represents a way to construct a larger vector space from two (or more) smaller vector spaces.

Definition 1.5 (Tensor product) Let V and W be Hilbert spaces (vector spaces with a scalar product) of dimension m and n with basis vectors $\{|v\rangle\}$ and $\{|w\rangle\}$, respectively. Then $V \otimes W$ is a Hilbert space of dimension $m \cdot n$, and a basis is spanned by $\{|v\rangle \otimes |w\rangle\}$, which is a set combining every basis vector of V with every basis vector of W.

Mathematical properties

- bilinearity $(z_1|v_1\rangle + z_2|v_2\rangle) \otimes |w\rangle = z_1|v_1\rangle \otimes |w\rangle + z_2|v_2\rangle \otimes |w\rangle$
- operators acting on the combined Hilbert space $A \otimes B$ act on the basis states as $(A \otimes B)(|v\rangle \otimes |w\rangle) = (A|v\rangle) \otimes (B|w\rangle)$
- any linear operator on $V \otimes W$ can be decomposed as $C = \sum_i c_i A_i \otimes B_i$
- the scalar product is inherited in the natural way; i.e., one has for $|a\rangle = \sum_{ij} a_{ij}|v_i\rangle \otimes |w_j\rangle$ and $|b\rangle = \sum_{k\ell} b_{k\ell}|v_k\rangle \otimes |w_\ell\rangle$ the scalar product $\langle a|b\rangle = \sum_{ijk\ell} a_{ij}^* b_{k\ell} \langle v_i|v_k\rangle \langle w_j|w_\ell\rangle = \sum_{ij} a_{ij}^* b_{ij}$

We note here that the basis vectors of the joint system are also often written as $|v\rangle \otimes |w\rangle = |vw\rangle$, where the order of v and w determines the subspace to which the quantum numbers are associated.

If more than just two vector spaces are combined to form a larger vector space, the definition of the tensor product may be applied recursively. As a consequence, the dimension of the joint vector space grows rapidly, as, e.g., exemplified by the case of a qubit: its Hilbert space is just spanned by two vectors $|0\rangle$ and $|1\rangle$. The joint Hilbert space of two qubits is spanned by the vectors $|0\rangle \otimes |0\rangle = |00\rangle$, $|0\rangle \otimes |1\rangle = |01\rangle$, $|1\rangle \otimes |0\rangle = |10\rangle$, and $|1\rangle \otimes |1\rangle = |11\rangle$, and is thus four dimensional. This can be readily scaled up: the dimension of the Hilbert space for three qubits is eight dimensional, and that for n qubits is 2^n dimensional. Eventually, this exponential growth of the Hilbert space dimension for composite quantum systems is at the heart of quantum computing.

Exercise 1.9 (Tensor products of operators) Let σ denote the Pauli matrices, i.e.,

$$\sigma^1 = \begin{pmatrix} 0 & +1 \\ +1 & 0 \end{pmatrix}, \qquad \sigma^2 = \begin{pmatrix} 0 & -i \\ +i & 0 \end{pmatrix}, \qquad \sigma^3 = \begin{pmatrix} +1 & 0 \\ 0 & -1 \end{pmatrix}. \quad (1.39)$$

Compute the trace of the operator

$$\Sigma = a\mathbf{1} \otimes \mathbf{1} + \sum_{i=1}^{3} \alpha_i \sigma^i \otimes \mathbf{1} + \sum_{j=1}^{3} \beta_j \mathbf{1} \otimes \sigma^j + \sum_{i,j=1}^{3} a_{ij} \sigma^i \otimes \sigma^j.$$

Since the scalar product in the subsystems is inherited by the scalar product of the composite system, this typically enables a convenient calculation of the trace—given a decomposition into only few tensor products. For example, one has for a single tensor product of two operators

$$
\begin{aligned}
\mathrm{Tr}\{A \otimes B\} &= \sum_{n_A, n_B} \langle n_A, n_B | A \otimes B | n_A, n_B \rangle \\
&= \left[\sum_{n_A} \langle n_A | A | n_A \rangle \right] \left[\sum_{n_B} \langle n_B | B | n_B \rangle \right] \\
&= \mathrm{Tr}_A\{A\} \mathrm{Tr}_B\{B\},
\end{aligned}
\tag{1.40}
$$

where $\mathrm{Tr}_{A/B}$ denote the trace in the Hilbert space of A and B, respectively. Since these traces only involve the summation over the degrees of freedom of a subsystem, they are also called partial traces. Such partial traces are of tremendous importance and will be discussed in the next section.

1.4.4 The Partial Trace

For composite systems, it is usually not necessary to keep all information of the complete system in the density matrix. Rather, one would like to have a density matrix that encodes all the information on a particular subsystem only. Obviously, the map $\rho \to \mathrm{Tr}_B\{\rho\}$ to such a reduced density matrix should leave all expectation values of observables acting on the considered subsystem only invariant, i.e.,

$$
\mathrm{Tr}\{A \otimes \mathbf{1}\rho\} = \mathrm{Tr}\{A\,\mathrm{Tr}_B\{\rho\}\}.
\tag{1.41}
$$

If this basic condition were not fulfilled, there would be no point in defining such a thing as a reduced density matrix: measurements would yield different results depending on the Hilbert space of the experimenter's choice.

Definition 1.6 (Partial trace) Let $|a_1\rangle$ and $|a_2\rangle$ be vectors of state space A and $|b_1\rangle$ and $|b_2\rangle$ vectors of state space B. Then, the partial trace over state space B is defined via

$$
\mathrm{Tr}_B\{|a_1\rangle\langle a_2| \otimes |b_1\rangle\langle b_2|\} = |a_1\rangle\langle a_2|\,\mathrm{Tr}\{|b_1\rangle\langle b_2|\}.
\tag{1.42}
$$

We note that whereas the trace mapped an operator to a number, the partial trace reduces operators to lower dimensional operators. The partial trace is linear, such that the partial trace of arbitrary operators is calculated similarly. By choosing the $|a_\alpha\rangle$ and $|b_\gamma\rangle$ as an orthonormal basis in the respective Hilbert space, one may therefore calculate the most general partial trace via

$$\mathrm{Tr_B}\{C\} = \mathrm{Tr_B}\left\{\sum_{\alpha\beta\gamma\delta} c_{\alpha\beta\gamma\delta}|a_\alpha\rangle\langle a_\beta| \otimes |b_\gamma\rangle\langle b_\delta|\right\}$$

$$= \sum_{\alpha\beta\gamma\delta} c_{\alpha\beta\gamma\delta}\,\mathrm{Tr_B}\left\{|a_\alpha\rangle\langle a_\beta| \otimes |b_\gamma\rangle\langle b_\delta|\right\}$$

$$= \sum_{\alpha\beta\gamma\delta} c_{\alpha\beta\gamma\delta}|a_\alpha\rangle\langle a_\beta|\,\mathrm{Tr}\left\{|b_\gamma\rangle\langle b_\delta|\right\}$$

$$= \sum_{\alpha\beta\gamma\delta} c_{\alpha\beta\gamma\delta}|a_\alpha\rangle\langle a_\beta|\sum_\epsilon \langle b_\epsilon|b_\gamma\rangle\langle b_\delta|b_\epsilon\rangle$$

$$= \sum_{\alpha\beta}\left[\sum_\gamma c_{\alpha\beta\gamma\gamma}\right]|a_\alpha\rangle\langle a_\beta|. \qquad (1.43)$$

Definition 1.6 is the only linear map that respects the invariance of expectation values [5].

Exercise 1.10 (Partial trace) Compute the partial trace $\rho_A = \mathrm{Tr_B}\{\rho_{AB}\}$ of a pure density matrix $\rho_{AB} = |\Psi\rangle\langle\Psi|$ in the bipartite state

$$|\Psi\rangle = \frac{1}{\sqrt{2}}\big(|01\rangle + |10\rangle\big) \equiv \frac{1}{\sqrt{2}}\big(|0\rangle\otimes|1\rangle + |1\rangle\otimes|0\rangle\big).$$

Show that ρ_A is no longer pure.

1.5 Lindblad Quantum Master Equation

Any dynamical evolution equation for the density matrix should preserve its interpretation as a density matrix. This implies that trace, hermiticity, and positivity or the initial condition must be preserved—at least in some approximate sense. By construction, the measurement postulate and unitary evolution preserve these properties. However, more general evolutions are conceivable as, e.g., exemplified by the Kraus map. If we constrain ourselves to master equations that are local in time and have constant coefficients, the most general evolution that preserves trace, self-adjointness, and positivity of the density matrix is given by a Lindblad form [19].

1.5.1 Representations

Definition 1.7 (Lindblad form)   A master equation of Lindblad form has the structure

$$\dot\rho = \mathcal{L}\rho = -\mathrm{i}[H,\rho] + \sum_{\alpha,\beta=1}^{N^2-1} \gamma_{\alpha\beta}\left(A_\alpha\rho A_\beta^\dagger - \frac{1}{2}\{A_\beta^\dagger A_\alpha, \rho\}\right), \qquad (1.44)$$

where the hermitian operator $H = H^\dagger$ can be interpreted as an effective Hamiltonian and the dampening matrix $\gamma_{\alpha\beta} = \gamma_{\beta\alpha}^*$ is a positive semidefinite matrix; i.e., it fulfills $\sum_{\alpha\beta} x_\alpha^* \gamma_{\alpha\beta} x_\beta \geq 0$ for all vectors x (or, equivalently, that all eigenvalues of $(\gamma_{\alpha\beta})$ are non-negative, $\lambda_i \geq 0$).

In the above definition, the commutator term with the effective Hamiltonian accounts for the unitary evolution, whereas the remaining terms are responsible for the nonunitary (dissipative) evolution. When derived from a microscopic model, the effective Hamiltonian need not coincide with the system Hamiltonian. This demonstrates that the interaction with a reservoir may also change the unitary part of the evolution.

Exercise 1.11 (Trace and hermiticity preservation by Lindblad forms) Show that the Lindblad form of master equation preserves trace and hermiticity of the density matrix.

The Lindblad-type master equation can be written in a simpler form. As the dampening matrix γ is hermitian, it can be diagonalized by a suitable unitary transformation U, such that $\sum_{\alpha\beta} U_{\alpha'\alpha} \gamma_{\alpha\beta} (U^\dagger)_{\beta\beta'} = \delta_{\alpha'\beta'} \gamma_{\alpha'}$ with $\gamma_\alpha \geq 0$ representing its non-negative eigenvalues. Using this unitary operation, a new set of operators can be defined via $A_\alpha = \sum_{\alpha'} U_{\alpha'\alpha} \bar{L}_{\alpha'}$. Inserting this decomposition in the master equation, we obtain

$$\dot{\rho} = -i[H, \rho] + \sum_{\alpha,\beta=1}^{N^2-1} \gamma_{\alpha\beta} \left(A_\alpha \rho A_\beta^\dagger - \frac{1}{2}\{A_\beta^\dagger A_\alpha, \rho\} \right)$$

$$= -i[H, \rho] + \sum_{\alpha',\beta'} \left[\sum_{\alpha\beta} \gamma_{\alpha\beta} U_{\alpha'\alpha} U_{\beta'\beta}^* \right] \left(\bar{L}_{\alpha'} \rho \bar{L}_{\beta'}^\dagger - \frac{1}{2}\{\bar{L}_{\beta'}^\dagger \bar{L}_{\alpha'}, \rho\} \right)$$

$$= -i[H, \rho] + \sum_\alpha \gamma_\alpha \left(\bar{L}_\alpha \rho \bar{L}_\alpha^\dagger - \frac{1}{2}\{\bar{L}_\alpha^\dagger \bar{L}_\alpha, \rho\} \right), \tag{1.45}$$

where γ_α denote the $N^2 - 1$ non-negative eigenvalues of the dampening matrix. Their positivity also allows us to absorb them into the Lindblad operators $L_\alpha \equiv \sqrt{\gamma_\alpha} \bar{L}_\alpha$ to yield the simplest representation of a Lindblad form,

$$\dot{\rho} = -i[H, \rho] + \sum_\alpha \left(L_\alpha \rho L_\alpha^\dagger - \frac{1}{2}\{L_\alpha^\dagger L_\alpha, \rho\} \right). \tag{1.46}$$

Evidently, the representation of a master equation is not unique. Any other unitary operation would lead to a different nondiagonal form which however describes the same master equation. In addition, we note here that the master equation is not only invariant to unitary transformations of the operators A_α, but in the diagonal representation also to inhomogeneous transformations of the form

$$L_\alpha \to L'_\alpha = L_\alpha + a_\alpha \mathbf{1},$$

$$H \to H' = H + \frac{1}{2i} \sum_\alpha (a_\alpha^* L_\alpha - a_\alpha L_\alpha^\dagger) + b\mathbf{1}, \qquad (1.47)$$

with complex numbers a_α and a real number b. The first of these equations can be exploited to choose the Lindblad operators L_α traceless, thereby fixing the numbers a_α, whereas b is fixed by gauging the energy of the Hamiltonian.

Exercise 1.12 (Shift invariance) Show the invariance of the diagonal representation of a Lindblad form master equation (1.46) with respect to the transformation (1.47).

1.5.2 Preservation of Positivity

Similar to the transformation into the interaction picture, one can eliminate the unitary evolution term by transforming Eq. (1.45) to a co-moving frame $\rho = e^{-iHt}\tilde{\rho}e^{+iHt}$. Then, the master equation assumes the form

$$\dot{\rho} = \sum_\alpha \gamma_\alpha \left(L_\alpha(t)\rho L_\alpha^\dagger(t) - \frac{1}{2}\{L_\alpha^\dagger(t)L_\alpha(t), \rho\} \right) \qquad (1.48)$$

with the transformed time-dependent operators $L_\alpha(t) = e^{+iHt}L_\alpha e^{-iHt}$. It is also clear that if the differential equation preserves positivity of the density matrix, then it would also do this for time-dependent rates γ_α. Define the operators with $K = N^2 - 1$

$$W_1(t) = \mathbf{1},$$

$$W_2(t) = \frac{1}{2\gamma} \sum_\alpha \gamma_\alpha(t) L_\alpha^\dagger(t)L_\alpha(t),$$

$$W_3(t) = L_1(t), \qquad (1.49)$$

$$\vdots$$

$$W_{K+2}(t) = L_K(t),$$

where $\gamma = \sum_\alpha \gamma_\alpha(t)$ has been introduced to render all W_i operators dimensionless. Discretizing the time derivative in Eq. (1.48), one transforms the differential equation for the density matrix into an iteration equation,

$$\rho(t + \Delta t) = \rho(t) + \Delta t \sum_\alpha \gamma_\alpha \left[L_\alpha(t)\rho(t)L_\alpha^\dagger(t) - \frac{1}{2}\{L_\alpha^\dagger(t)L_\alpha(t), \rho(t)\} \right]$$

$$= \sum_{\alpha\beta} w_{\alpha\beta}(t) W_\alpha(t)\rho(t) W_\beta^\dagger(t), \tag{1.50}$$

where the dimensionless $w_{\alpha\beta}$ matrix assumes the block form

$$w(t) = \begin{pmatrix} 1 & -\Delta t\gamma & 0 & \cdots & 0 \\ -\Delta t\gamma & 0 & 0 & \cdots & 0 \\ \hline 0 & 0 & \Delta t\gamma_1(t) & & \\ \vdots & \vdots & & \ddots & \\ 0 & 0 & & & \Delta t\gamma_K(t) \end{pmatrix}, \tag{1.51}$$

which makes it particularly easy to diagonalize; the lower right block is already diagonal and the eigenvalues of the upper 2 by 2 block may be directly obtained by solving for the roots of the characteristic polynomial $\lambda^2 - \lambda - (\gamma \Delta t)^2 = 0$. Again, we introduce the corresponding unitary transformation $\tilde{W}_\alpha(t) = \sum_{\alpha'} u_{\alpha'\alpha}(t)W_{\alpha'}(t)$ to find that

$$\rho(t + \Delta t) = \sum_\alpha w_\alpha(t)\tilde{W}_\alpha(t)\rho(t)\tilde{W}_\alpha^\dagger(t) \tag{1.52}$$

with $w_\alpha(t)$ denoting the eigenvalues of the matrix (1.51) and in particular the only negative eigenvalue being given by $w_1(t) = \frac{1}{2}(1 - \sqrt{1 + 4(\gamma \Delta t)^2})$. Now, we use the spectral decomposition of the density matrix at time t, $\rho(t) = \sum_a P_a(t)|\Psi_a(t)\rangle\langle\Psi_a(t)|$, to demonstrate approximate positivity of the density matrix at time $t + \Delta t$:

$$\langle\Phi|\rho(t + \Delta t)|\Phi\rangle = \sum_{\alpha,a} w_\alpha(t) P_a(t) \big|\langle\Phi|\tilde{W}_\alpha(t)|\Psi_a(t)\rangle\big|^2$$

$$\geq \frac{1}{2}\left(1 - \sqrt{1 + 4(\gamma \Delta t)^2}\right) \sum_a P_a(t) \big|\langle\Phi|\tilde{W}_1(t)|\Psi_a(t)\rangle\big|^2$$

$$\geq -(\gamma \Delta t)^2 \sum_a P_a(t) \big|\langle\Phi|\tilde{W}_1(t)|\Psi_a(t)\rangle\big|^2 \overset{\Delta t \to 0}{\longrightarrow} 0, \tag{1.53}$$

such that the violation of positivity vanishes faster than the discretization width as Δt goes to zero. This limit just yields the differential equation of the Lindblad form master equation, which shows that the latter preserves positivity. It should be noted however that numerical solutions of the Lindblad master equation using a forward-time discretization may yield negative probabilities if the time step Δt is chosen too large.

1.5.3 Rate Equation Representation

Since Eq. (1.45) at all times preserves hermiticity of the density matrix, it can always be diagonalized by a unitary transformation $\rho_D(t) = U(t)\rho(t)U^\dagger(t)$. Inserting this transformation in the master equation (1.46) yields

$$\dot{\rho}_D = \dot{U}U^\dagger\rho_D + U\dot{\rho}U^\dagger + \rho_D U\dot{U}^\dagger = -\mathrm{i}[\mathrm{i}\dot{U}U^\dagger, \rho_D] + U\dot{\rho}U^\dagger$$

$$= -\mathrm{i}[\boldsymbol{H}(t) + \boldsymbol{H}_{\mathrm{eff}}(t), \rho_D]$$

$$+ \sum_\alpha \left[\boldsymbol{L}_\alpha(t)\rho_D\boldsymbol{L}_\alpha^\dagger(t) - \frac{1}{2}\{\boldsymbol{L}_\alpha^\dagger(t)\boldsymbol{L}_\alpha(t), \rho_D\}\right], \qquad (1.54)$$

with transformed Lindblad operators $\boldsymbol{L}_\alpha(t) = U(t)L_\alpha U^\dagger(t)$ and the effective Hamiltonian $\boldsymbol{H}_{\mathrm{eff}}(t) = \mathrm{i}\dot{U}(t)U^\dagger(t)$.

Exercise 1.13 (Hermiticity of effective Hamiltonian) Show that the effective Hamiltonian $\boldsymbol{H}_{\mathrm{eff}}(t) = \mathrm{i}\dot{U}(t)U^\dagger(t)$ is hermitian.

Now using the fact that in the time-dependent basis ρ_D is diagonal, $\rho_D = \sum_a \rho_{aa}(t)|a(t)\rangle\langle a(t)|$, we obtain

$$\dot{\rho}_{aa} = \sum_\alpha \sum_b |\langle a|\boldsymbol{L}_\alpha(t)|b\rangle|^2 \rho_{bb} - \sum_\alpha \langle a|\boldsymbol{L}_\alpha^\dagger(t)\boldsymbol{L}_\alpha(t)|a\rangle \rho_{aa}, \qquad (1.55)$$

which has the structure of a rate equation with positive but time-dependent rates [20]. Unfortunately, to obtain such a rate equation, one first has to diagonalize the time-dependent solution of Eq. (1.46), i.e., to solve the complicated master equation beforehand. It is therefore not very practical in most cases, unless one is given a rate equation from the start. Nevertheless, it shows that rate equations—if set up in the correct basis—can yield a quite general description of master equation dynamics. The basis within which the long-term density matrix becomes diagonal is also called a pointer basis.

1.5.4 Examples

1.5.4.1 Cavity in a Thermal Bath

Consider the Lindblad form master equation

$$\dot{\rho}_S = -\mathrm{i}[\Omega a^\dagger a, \rho_S] + \gamma(1 + n_B)\left[a\rho_S a^\dagger - \frac{1}{2}a^\dagger a\rho_S - \frac{1}{2}\rho_S a^\dagger a\right]$$

$$+ \gamma n_B\left[a^\dagger \rho_S a - \frac{1}{2}aa^\dagger \rho_S - \frac{1}{2}\rho_S aa^\dagger\right]. \qquad (1.56)$$

Here a and $a^\dagger$ are bosonic annihilation and creation operators, respectively, fulfilling the bosonic commutation relations $[a, a^\dagger] = \mathbf{1}$. The dampening matrix is given by the Bose–Einstein bath occupation $n_B = [e^{\beta\Omega} - 1]^{-1}$ evaluated at cavity frequency Ω and a bare emission and absorption rate $\gamma > 0$. In Fock space representation, these operators act as $a^\dagger |n\rangle = \sqrt{n+1}|n+1\rangle$ and $a|n\rangle = \sqrt{n}|n-1\rangle$ (where $0 \leq n < \infty$), such that the above master equation couples only the diagonals of the density matrix $\rho_n = \langle n|\rho_S|n\rangle$ to each other,

$$\dot{\rho}_n = \gamma(1 + n_B)\big[(n + 1)\rho_{n+1} - n\rho_n\big] + \gamma n_B\big[n\rho_{n-1} - (n + 1)\rho_n\big]$$

$$= \gamma n_B n\rho_{n-1} - \gamma\big[n + (2n + 1)n_B\big]\rho_n + \gamma(1 + n_B)(n + 1)\rho_{n+1}, \quad (1.57)$$

in a tridiagonal form. That makes it particularly easy to calculate the stationary state of the populations recursively, since the boundary solution $n_B\bar{\rho}_0 = (1 + n_B)\bar{\rho}_1$ implies for all n the relation

$$\frac{\bar{\rho}_{n+1}}{\bar{\rho}_n} = \frac{n_B}{1 + n_B} = e^{-\beta\Omega}. \quad (1.58)$$

Consequently, the stationary populations are consistent with a thermalized Gibbs state

$$\bar{\rho} = \frac{e^{-\beta\Omega a^\dagger a}}{\mathrm{Tr}\{e^{-\beta\Omega a^\dagger a}\}} \quad (1.59)$$

with the inverse reservoir temperature β. Such a Gibbs state however does not have coherences in the Fock space basis. To investigate their evolution, we calculate the time derivative of $\rho_{nm} = \langle n|\rho_S|m\rangle$,

$$\dot{\rho}_{nm} = \left[-i\Omega(n - m) - \gamma(1 + n_B)\frac{n + m}{2} - \gamma n_B\frac{n + 1 + m + 1}{2}\right]\rho_{nm}$$

$$+ \gamma(1 + n_B)\sqrt{(n + 1)(m + 1)}\rho_{n+1,m+1} + \gamma n_B\sqrt{nm}\rho_{n-1,m-1}, \quad (1.60)$$

which, when $n \neq m$, shows that the coherences do not formally depend on the dynamics of the populations. However, they couple strongly to other coherences. In particular, we observe that coherences $\rho_n^\Delta = \rho_{n,n+\Delta}$ with integer $\Delta \neq 0$ and $\Delta \geq -n$ couple only to coherences with the same difference $\rho_{n\pm1}^\Delta$:

$$\dot{\rho}_n^\Delta = \big[+i\Omega\Delta - \gamma(1 + n_B)(n + \Delta/2) - \gamma n_B(n + 1 + \Delta/2)\big]\rho_n^\Delta$$

$$+ \gamma(1 + n_B)\sqrt{(n + 1)(n + 1 + \Delta)}\rho_{n+1}^\Delta + \gamma n_B\sqrt{n(n + \Delta)}\rho_{n-1}^\Delta, \quad (1.61)$$

which also corresponds to a tridiagonal system for each fixed difference Δ. For each Δ it is straightforward to see that $\bar{\rho}_n^\Delta = 0$ is a stationary solution.

Exercise 1.14 (Moments) Calculate the expectation value of the number operator $n = a^\dagger a$ and its square $n^2 = a^\dagger a a^\dagger a$ in the stationary state of the master equation (1.56).

1.5.4.2 Driven Cavity with Losses

When the cavity is driven with a laser and simultaneously coupled to a vacuum bath $n_B = 0$, we obtain the master equation

$$\dot{\rho}_S = -i\left[\Omega a^\dagger a + \frac{P}{2}e^{+i\omega t}a + \frac{P^*}{2}e^{-i\omega t}a^\dagger, \rho_S\right]$$
$$+ \gamma\left[a\rho_S a^\dagger - \frac{1}{2}a^\dagger a\rho_S - \frac{1}{2}\rho_S a^\dagger a\right] \qquad (1.62)$$

with the laser frequency ω and amplitude P. The transformation $\rho = e^{+i\omega a^\dagger a t} \times \rho_S e^{-i\omega a^\dagger a t}$ maps to a time-independent master equation,

$$\dot{\rho} = -i\left[(\Omega - \omega)a^\dagger a + \frac{P}{2}a + \frac{P^*}{2}a^\dagger, \rho\right] + \gamma\left[a\rho a^\dagger - \frac{1}{2}a^\dagger a\rho - \frac{1}{2}\rho a^\dagger a\right]. \qquad (1.63)$$

This equation obviously couples coherences and populations in the Fock space representation. Therefore, it does not assume a simple rate equation form in this basis. Nevertheless, a solution of the resulting equation of motion can be found for particular operators.

Exercise 1.15 (Coherent state) Using the driven cavity master equation, show that the stationary expectation value of the cavity occupation fulfills

$$\lim_{t\to\infty}\langle a^\dagger a\rangle = \frac{|P|^2}{\gamma^2 + 4(\Omega - \omega)^2}.$$

1.6 Superoperator Notation

The Lindblad master equation may be a bit impractical for calculations, as one is often more used to the solution of first-order differential equations that are written as $\dot{v} = Av$, where A is a matrix and v is a vector. Since the Lindblad equation is linear in the density matrix ρ, one can easily convert it into such a form, where one writes $\dot{\rho} = \mathscr{L}\rho$. In this representation, $\mathscr{L}$ is a matrix, and the density matrix becomes a density vector. Conventionally, the mapping to a density vector is performed by first placing d populations and then the $d(d-1)$ coherences:

$$\rho = \begin{pmatrix} \rho_{11} & \cdots & \rho_{1N} \\ \vdots & & \vdots \\ \rho_{N1} & \cdots & \rho_{NN} \end{pmatrix} \quad \Leftrightarrow \quad \|\rho\rangle\rangle = \begin{pmatrix} \rho_{11} \\ \vdots \\ \rho_{NN} \\ \rho_{12} \\ \rho_{21} \\ \vdots \\ \rho_{N-1,N} \\ \rho_{N,N-1} \end{pmatrix}. \tag{1.64}$$

A master equation can now be written as

$$\dot{\rho} = \mathscr{L}\rho, \tag{1.65}$$

where the superoperator corresponding to the Lindblad form (1.46) acts like an ordinary operator on the density vector. In this representation, the trace of a density matrix corresponds to multiplication with the vector

$$\langle\langle 0\| = (\underbrace{1,\ldots,1}_{N\times}, \underbrace{0,\ldots,0}_{N(N-1)\times}), \tag{1.66}$$

i.e., $\mathrm{Tr}\{\rho\} = \langle\langle 0\|\rho\rangle\rangle$. Thus, when the Hilbert space dimension of the quantum system is d, ρ_S is a vector of dimension d^2, and the superoperator $\mathscr{L}$ is represented by a $d^2 \times d^2$ matrix. At first sight, such a representation does not seem very efficient. However, for many specific cases, the structure of the superoperator $\mathscr{L}$ may directly allow for a simplified treatment. If for example it has block structure—as will be the case in the quantum optical master equation when the system has no degeneracies—one may treat the blocks separately, which is routinely done. To be more specific, the mapping can generally be performed as

$$\dot{\rho}_{ij} = -\mathrm{i}\langle i| \left[H, \sum_{kl} \rho_{kl}|k\rangle\langle l| \right] |j\rangle$$

$$+ \sum_{\alpha\beta} \gamma_{\alpha\beta} \left[\langle i|A_\alpha \sum_{kl} \rho_{kl}|k\rangle\langle l|A_\beta^\dagger|j\rangle - \frac{1}{2}\langle i| \left\{ A_\beta^\dagger A_\alpha, \sum_{kl} \rho_{kl}|k\rangle\langle l| \right\} |j\rangle \right]$$

$$= \sum_{kl} \mathscr{L}_{ij,kl}\rho_{kl}. \tag{1.67}$$

Similarly, we can also transform linear operators into superoperators. However, we must specify on which side of the density matrix the operator is supposed to act.

As an example, we consider the Liouvillian

$$L[\rho] = -\mathrm{i}\left[\Omega\sigma^z, \rho\right] + \gamma\left[\sigma^-\rho\sigma^+ - \frac{1}{2}\{\sigma^+\sigma^-, \rho\}\right] \tag{1.68}$$

with $\sigma^\pm = \frac{1}{2}(\sigma^x \pm i\sigma^y)$ in the eigenbasis of $\sigma^z|e\rangle = |e\rangle$ and $\sigma^z|g\rangle = -|g\rangle$, where we have $\sigma^+|g\rangle = |e\rangle$ and $\sigma^-|e\rangle = |g\rangle$. From the master equation, we obtain

$$\dot{\rho}_{ee} = -\gamma\rho_{ee}, \qquad \dot{\rho}_{gg} = +\gamma\rho_{ee},$$

$$\dot{\rho}_{eg} = \left(-\frac{\gamma}{2} - 2i\Omega\right)\rho_{eg}, \qquad \dot{\rho}_{ge} = \left(-\frac{\gamma}{2} + 2i\Omega\right)\rho_{ge}, \tag{1.69}$$

such that when we arrange the matrix elements in a vector $\rho = (\rho_{gg}, \rho_{ee}, \rho_{ge}, \rho_{eg})^T$, the master equation reads

$$\begin{pmatrix} \dot{\rho}_{gg} \\ \dot{\rho}_{ee} \\ \dot{\rho}_{ge} \\ \dot{\rho}_{eg} \end{pmatrix} = \begin{pmatrix} 0 & +\gamma & 0 & 0 \\ 0 & -\gamma & 0 & 0 \\ 0 & 0 & -\frac{\gamma}{2} + 2i\Omega & 0 \\ 0 & 0 & 0 & -\frac{\gamma}{2} - 2i\Omega \end{pmatrix} \begin{pmatrix} \rho_{gg} \\ \rho_{ee} \\ \rho_{ge} \\ \rho_{eg} \end{pmatrix}. \tag{1.70}$$

We note that populations and coherences evolve apparently independently. Note however that the Lindblad form nevertheless ensures a positive density matrix—with valid initial conditions provided.

Exercise 1.16 (Preservation of Positivity) Show that the superoperator in Eq. (1.70) preserves positivity of the density matrix provided that initial positivity ($-1/4 \leq |\rho_{ge}^0|^2 - \rho_{gg}^0\rho_{ee}^0 \leq 0$) is given.

Furthermore, note that we do not need to exponentiate a matrix to solve Eq. (1.70): its special structure makes it possible to solve for ρ_{eg} and ρ_{ge}, and ρ_{ee} independently. The equation for ρ_{gg} does depend on the result for ρ_{ee}; however, we may readily obtain the solution by exploiting trace conservation $\rho_{gg} = 1 - \rho_{ee}$. It is of course also possible to represent ordinary operators as superoperators. This however requires one to specify on which side the operator is acting; for example, one has

$$\begin{pmatrix} 0 & 0 \\ 0 & 1 \end{pmatrix}\begin{pmatrix} \rho_{gg} & \rho_{ge} \\ \rho_{eg} & \rho_{ee} \end{pmatrix} \hat{=} \begin{pmatrix} 0 & 0 & 0 & 0 \\ 0 & 1 & 0 & 0 \\ 0 & 0 & 0 & 0 \\ 0 & 0 & 0 & 1 \end{pmatrix}\begin{pmatrix} \rho_{gg} \\ \rho_{ee} \\ \rho_{ge} \\ \rho_{eg} \end{pmatrix},$$

$$\begin{pmatrix} \rho_{gg} & \rho_{ge} \\ \rho_{eg} & \rho_{ee} \end{pmatrix}\begin{pmatrix} 0 & 0 \\ 0 & 1 \end{pmatrix} \hat{=} \begin{pmatrix} 0 & 0 & 0 & 0 \\ 0 & 1 & 0 & 0 \\ 0 & 0 & 1 & 0 \\ 0 & 0 & 0 & 0 \end{pmatrix}\begin{pmatrix} \rho_{gg} \\ \rho_{ee} \\ \rho_{ge} \\ \rho_{eg} \end{pmatrix}. \tag{1.71}$$

Exercise 1.17 (Expectation values from superoperators) Show that for a Liouvillian superoperator connecting N populations (diagonal entries) with M coherences (off-diagonal entries) acting on the density matrix $\rho(t) = (P_1, \ldots, P_N, C_1, \ldots, C_M)^T$, the trace in the expectation value of an operator can be mapped to the matrix element

$$\langle A(t) \rangle = (\underbrace{1, \dots, 1}_{N \times}, \underbrace{0, \dots, 0}_{M \times}) \cdot \mathscr{A} \cdot \rho(t),$$

where the matrix $\mathscr{A}$ is the superoperator corresponding to multiplication with A from the left.

References

1. P.W. Shor, Polynomial-time algorithms for prime factorization and discrete logarithms on a quantum computer. SIAM J. Comput. **26**, 1484 (1997)
2. L.K. Grover, Fixed-point quantum search. Phys. Rev. Lett. **95**, 150501 (2005)
3. R.P. Feynman, Simulating physics with computers. Int. J. Theor. Phys. **21**, 467 (1982)
4. S. Lloyd, Universal quantum simulators. Science **273**, 1073 (1996)
5. M.A. Nielsen, I.L. Chuang, *Quantum Computation and Quantum Information* (Cambridge University Press, Cambridge, 2000)
6. H.-P. Breuer, F. Petruccione, *The Theory of Open Quantum Systems* (Oxford University Press, Oxford, 2002)
7. M. Schlosshauer, *Decoherence and the Quantum-to-Classical Transition* (Springer, Berlin, 2007)
8. L.M.K. Vandersypen, M. Steffen, G. Breyta, C.S. Yannoni, M.H. Sherwood, I.L. Chuang, Experimental realization of Shor's quantum factoring algorithm using nuclear magnetic resonance. Nature **414**, 883 (2001)
9. J.I. Cirac, P. Zoller, New frontiers in quantum information with atoms and ions. Phys. Today **57**, 38 (2004)
10. J. Du, N. Xu, X. Peng, P. Wang, S. Wu, D. Lu, NMR implementation of a molecular hydrogen quantum simulation with adiabatic state preparation. Phys. Rev. Lett. **104**, 030502 (2010)
11. R. Coldea, D.A. Tennant, E.M. Wheeler, E. Wawrzynska, D. Prabhakaran, M. Telling, K. Habicht, P. Smeibidl, K. Kiefer, Quantum criticality in an Ising chain: experimental evidence for emergent e_8 symmetry. Science **327**, 177 (2010)
12. R. Sánchez, M. Buttiker, Optimal energy quanta to current conversion. Physical Review B **83**, 085428 (2011)
13. N. Linden, S. Popescu, P. Skrzypczyk, How small can thermal machines be? The smallest possible refrigerator. Phys. Rev. Lett. **105**, 130401 (2010)
14. V.M. Bastidas, C. Emary, G. Schaller, A. Gómez-León, G. Platero, T. Brandes, Floquet topological quantum phase transitions in the transverse Wen-plaquette model (2013). arXiv:1302.0781
15. J. Dziarmaga, Dynamics of a quantum phase transition: exact solution of the quantum Ising model. Phys. Rev. Lett. **95**, 245701 (2005)
16. E. Barnes, S. Das Sarma, Analytically solvable driven time-dependent two-level quantum systems. Phys. Rev. Lett. **109**, 060401 (2012)
17. N. Linden, S. Popescu, A.J. Short, A. Winter, Quantum mechanical evolution towards thermal equilibrium. Phys. Rev. E **79**, 061103 (2009)
18. K. Kraus, *Effects and Operations: Fundamental Notions of Quantum Theory* (Springer, Berlin, 1983)
19. G. Lindblad, On the generators of quantum dynamical semigroups. Commun. Math. Phys. **48**, 119 (1976)
20. M. Esposito, S. Mukamel, Fluctuation theorems for quantum master equations. Phys. Rev. E **73**, 046129 (2006)

Chapter 2
Microscopic Derivation

Abstract In this chapter, we provide methods of deriving evolution equations for the density matrix from a microscopic model defined by system, reservoir, and interaction Hamiltonians. Since all methods assume that the interaction Hamiltonian can be decomposed into tensor products of system and bath operators, we first demonstrate how to convert fermionic tunnel couplings into these representations. Then, we introduce a Kraus-type map for the density matrix that is valid for short times and/or weak couplings. The corresponding master equation is introduced via a coarse-graining approach, which however for large coarse-graining times reproduces the quantum optical master equation, valid in the weak coupling limit. Finally, we discuss important properties of the quantum optical master equation and the singular coupling limit. Here, it is always assumed that the interaction between system and reservoir does not change the state of the reservoir. The examples in later chapters will often refer to the definitions in this chapter, which may therefore also be used as a reference.

Given a microscopic model, the actual derivation of a master equation may be quite challenging and can be an art of its own for some parameter regimes. However, there exist quite well-known and model-independent limits where the road map to the master equation is well documented. The most important one is the weak coupling limit [1]. Often, such a microscopic derivation is rewarding from a conceptual point of view, since one gets a thermodynamic interpretation for free.

Before we start discussing the derivations, we make a technical remark: to perform a microscopic derivation, it is required to perform a partial trace over all degrees of freedom that are not considered as belonging to the system. To perform this trace in a convenient way, we assume a decomposition of the interaction Hamiltonian in terms of a tensor product,

$$H_{\mathrm{I}} = \sum_\alpha A_\alpha \otimes B_\alpha, \tag{2.1}$$

where A_α are system operators and B_α are bath operators, respectively. This tensor product decomposition implies that the commutator of system and bath operators vanishes, $[A_\alpha, B_\beta] = 0$. In fermionic tunneling terms, the standard representation

G. Schaller, *Open Quantum Systems Far from Equilibrium*, Lecture Notes in Physics 881,   27
DOI 10.1007/978-3-319-03877-3_2,
© Springer International Publishing Switzerland 2014

has however anti-commuting operators associated with system and bath, respectively. Therefore, we first demonstrate below how such terms can be mapped to a tensor product representation.

2.1 Tensor Product Representation of Fermionic Tunnel Couplings

The representation (2.1) is not compatible with a fermionic tunneling Hamiltonian of, e.g., the form

$$H_{\mathrm{I}} = d \sum_k t_k c_k^\dagger + \sum_k t_k^* c_k d = d \sum_k t_k c_k^\dagger - d^\dagger \sum_k t_k^* c_k, \qquad (2.2)$$

where the sign arises because the anti-commutator of system and bath operators vanishes, $\{d, c_k^\dagger\} = 0$. It is however possible to map fermionic operators to spin operators via the Jordan–Wigner transformation. This transformation decomposes the fermionic operators in terms of Pauli matrices acting on different spins

$$d = \sigma^- \otimes \mathbf{1} \otimes \mathbf{1} \otimes \cdots \otimes \mathbf{1},$$

$$c_k = \sigma^z \otimes \underbrace{\sigma^z \otimes \cdots \otimes \sigma^z}_{k-1} \otimes \sigma^- \otimes \mathbf{1} \otimes \cdots \otimes \mathbf{1}, \qquad (2.3)$$

to map to a tensor product decomposition of the interaction Hamiltonian, where $\sigma^\pm = \frac{1}{2}[\sigma^x \pm i\sigma^y]$. The remaining operators follow from $(\sigma^+)^\dagger = \sigma^-$ and vice versa. This decomposition automatically obeys the fermionic anti-commutation relations such as $\{c_k, d^\dagger\} = 0$ and may therefore also be used to create a fermionic operator basis with computer algebra programs.

Exercise 2.1 (Jordan–Wigner transform) Show that for fermions distributed on N sites, the decomposition

$$c_i = \underbrace{\sigma^z \otimes \cdots \otimes \sigma^z}_{i-1} \otimes \sigma^- \otimes \underbrace{\mathbf{1} \otimes \cdots \otimes \mathbf{1}}_{N-i}$$

preserves the fermionic anti-commutation relations

$$\{c_i, c_j\} = \mathbf{0} = \{c_i^\dagger, c_j^\dagger\}, \qquad \{c_i, c_j^\dagger\} = \delta_{ij} \mathbf{1}.$$

Show also that the fermionic Fock space basis $c_i^\dagger c_i |n_1, \ldots, n_N\rangle = n_i |n_1, \ldots, n_N\rangle$ obeys $\sigma_i^z |n_1, \ldots, n_N\rangle = (-1)^{n_i+1} |n_1, \ldots, n_N\rangle$.

Using the fact that for the Pauli matrices (1.39) one has $\sigma^- \sigma^z = \sigma^-$ and $\sigma^+ \sigma^z = -\sigma^+$, the interaction Hamiltonian becomes (omitting all identity operators)

$$H_I = \sigma^- \otimes \sum_k t_k \underbrace{\sigma^z \otimes \cdots \otimes \sigma^z}_{k-1} \otimes \sigma^+ + \sigma^+ \otimes \sum_k t_k^* \underbrace{\sigma^z \otimes \cdots \otimes \sigma^z}_{k-1} \otimes \sigma^-,$$

$$(2.4)$$

which is compatible with Eq. (2.1).

Calculations with such lengthy spin operators may be inconvenient, such that one may reintroduce fermionic operators defined separately on system and bath Hilbert spaces, respectively,

$$\tilde{d} = \sigma^-, \qquad \tilde{c}_k = \underbrace{\sigma^z \otimes \cdots \otimes \sigma^z}_{k-1} \otimes \sigma^- \otimes \mathbf{1} \otimes \cdots \otimes \mathbf{1}, \qquad (2.5)$$

such that the interaction Hamiltonian in terms of these operators becomes

$$H_I = \tilde{d} \otimes \sum_k t_{kL} \tilde{c}_k^\dagger + \tilde{d}^\dagger \otimes \sum_k t_{kL}^* \tilde{c}_k, \qquad (2.6)$$

which of course also respects the required tensor decomposition (2.1). In the fermionic models that follow, we will implicitly assume such mappings. It is however also possible to perform specific calculations with the original fermionic operators yielding the same result.

2.2 A Mapping for Short Times or Weak Couplings

We assume a decomposition of the Hamiltonian

$$H = H_S + H_I + H_B \qquad (2.7)$$

into system, an interaction of type (2.1), and bath Hamiltonians, respectively. Aiming at a perturbative treatment in the interaction, we transform to the interaction picture

$$\rho(t) = e^{+i(H_S+H_B)t} \rho(t) e^{-i(H_S+H_B)t} = e^{+iH_St} e^{+iH_Bt} \rho(t) e^{-iH_Bt} e^{-iH_St}, \quad (2.8)$$

where the von Neumann equation in the interaction picture

$$\dot{\rho} = -i\big[H_I(t), \rho(t)\big] \qquad (2.9)$$

just contains the time-dependent interaction Hamiltonian. To simplify the bookkeeping of terms and notation, we introduce the dimensionless perturbation parameter λ by transforming $H_I(t) \to \lambda H_I(t)$. In the end, we will replace $\lambda \to 1$. For factorizing initial density matrices, the von Neumann equation is formally solved by $U(t)\rho_S^0 \otimes \bar{\rho}_B U^\dagger(t)$, where the time evolution operator

$$U(t) = \hat{\tau} \exp\left\{ -i\lambda \int_0^t H_I(t')\,dt' \right\} \qquad (2.10)$$

obeys the evolution equation

$$\dot{U} = -i\lambda H_I(t) U(t), \tag{2.11}$$

which defines the time-ordering operator $\hat{\tau}$. Formally integrating this equation with the evident initial condition $U(0) = 1$ yields

$$U(t) = 1 - i\lambda \int_0^t H_I(t') U(t') \, dt'$$

$$= 1 - i\lambda \int_0^t H_I(t') \, dt' - \lambda^2 \int_0^t dt' H_I(t') \left[\int_0^{t'} dt'' H_I(t'') U(t'') \right]$$

$$= 1 - i\lambda \int_0^t H_I(t') \, dt' - \lambda^2 \int_0^t dt_1 \, dt_2 \, H_I(t_1) H_I(t_2) \Theta(t_1 - t_2)$$

$$+ \mathscr{O}\{\lambda^3\}, \tag{2.12}$$

where the occurrence of the Heaviside function $\Theta(x)$ is a consequence of time ordering. For the hermitian conjugate operator we obtain

$$U^\dagger(t) \approx 1 + i\lambda \int_0^t H_I(t') \, dt' - \lambda^2 \int_0^t dt_1 \, dt_2 \, H_I(t_1) H_I(t_2) \Theta(t_2 - t_1)$$

$$+ \mathscr{O}\{\lambda^3\}. \tag{2.13}$$

To keep the discussion at a moderate level, we assume $\mathrm{Tr}\{B_\alpha \bar{\rho}_B\} = 0$. Though for many conceivable models this is fulfilled, it is a priori not the general case. However, it can be shown that this case can always be achieved by suitable transformation of the system and interaction Hamiltonians.

Exercise 2.2 (Transforming the coupling operators) Given an interaction Hamiltonian $H_I = \sum_\alpha A_\alpha \otimes B_\alpha$ where $\langle B_\alpha \rangle \neq 0$, show that there exists a simple transformation $B_\alpha \rightarrow B'_\alpha$ and $H_S \rightarrow H'_S$ which obeys $\langle B'_\alpha \rangle = 0$. Find B'_α and H'_S.

The exact solution is approximated by

$$\rho_S(t) = \mathrm{Tr}_B \left\{ \left[1 - i\lambda \int_0^t H_I(t_1) \, dt_1 \right. \right.$$

$$\left. - \lambda^2 \int_0^t dt_1 \, dt_2 \, H_I(t_1) H_I(t_2) \Theta(t_1 - t_2) \right] \rho_S^0 \otimes \bar{\rho}_B$$

$$\times \left[1 + i\lambda \int_0^t H_I(t_1) \, dt_1 - \lambda^2 \int_0^t dt_1 \, dt_2 \, H_I(t_1) H_I(t_2) \Theta(t_2 - t_1) \right] \right\}$$

$$+ \mathscr{O}\{\lambda^3\}$$

$$= \rho_S^0 - i\lambda \, \mathrm{Tr}_B \left\{ \left[\int_0^t H_I(t_1) \, dt_1, \rho_S^0 \otimes \bar{\rho}_B \right] \right\}$$

$$+ \lambda^2 \mathrm{Tr_B} \left\{ \int_0^t dt_1 \int_0^t dt_2 \, \boldsymbol{H}_I(t_1) \rho_S^0 \otimes \bar{\rho}_B \boldsymbol{H}_I(t_2) \right\}$$

$$- \lambda^2 \int_0^t dt_1 \, dt_2 \, \mathrm{Tr_B} \left\{ \Theta(t_1 - t_2) \boldsymbol{H}_I(t_1) \boldsymbol{H}_I(t_2) \rho_S^0 \otimes \bar{\rho}_B \right.$$

$$\left. + \Theta(t_2 - t_1) \rho_S^0 \otimes \bar{\rho}_B \boldsymbol{H}_I(t_1) \boldsymbol{H}_I(t_2) \right\}$$

$$+ \mathcal{O}\{\lambda^3\}. \tag{2.14}$$

We introduce the bath correlation functions with two time arguments:

$$C_{\alpha\beta}(t_1, t_2) = \mathrm{Tr}\{\boldsymbol{B}_\alpha(t_1) \boldsymbol{B}_\beta(t_2) \bar{\rho}_B\}, \tag{2.15}$$

such that we have, using that $\mathrm{Tr_B}\{\boldsymbol{B}_\alpha(t) \bar{\rho}_B\} = 0$, the equation

$$\boldsymbol{\rho}_S(t) = \rho_S^0 + \lambda^2 \sum_{\alpha\beta} \int_0^t dt_1 \int_0^t dt_2 \, C_{\alpha\beta}(t_1, t_2) \big[\boldsymbol{A}_\beta(t_2) \rho_S^0 \boldsymbol{A}_\alpha(t_1)$$

$$- \Theta(t_1 - t_2) \boldsymbol{A}_\alpha(t_1) \boldsymbol{A}_\beta(t_2) \rho_S^0 - \Theta(t_2 - t_1) \rho_S^0 \boldsymbol{A}_\alpha(t_1) \boldsymbol{A}_\beta(t_2) \big]$$

$$+ \mathcal{O}\{\lambda^3\}. \tag{2.16}$$

We now aim to identify this expression with a positivity-preserving Kraus map. To do so, we first insert a time-independent basis and use $\Theta(x) = \frac{1}{2}[1 + \mathrm{sgn}(x)]$ to obtain

$$\boldsymbol{\rho}_S(t) = \rho_S^0 + \lambda^2 \sum_{\alpha\beta} \int_0^t dt_1 \int_0^t dt_2 \, C_{\alpha\beta}(t_1, t_2)$$

$$\times \sum_{abcd} \langle a | \boldsymbol{A}_\beta(t_2) | b \rangle \langle c | \boldsymbol{A}_\alpha^\dagger(t_1) | d \rangle^* \big[|a\rangle\langle b| \rho_S^0 \big(|c\rangle\langle d| \big)^\dagger$$

$$- \Theta(t_1 - t_2) \big(|c\rangle\langle d| \big)^\dagger |a\rangle\langle b| \rho_S^0 - \Theta(t_2 - t_1) \rho_S^0 \big(|c\rangle\langle d| \big)^\dagger |a\rangle\langle b| \big] + \mathcal{O}\{\lambda^3\}$$

$$\equiv \rho_S^0 + \lambda^2 \sum_{abcd} \gamma_{ab,cd}(t) \Big[L_{ab} \rho_S^0 L_{cd}^\dagger - \frac{1}{2} \big\{ L_{cd}^\dagger L_{ab}, \rho_S^0 \big\} \Big]$$

$$- \mathrm{i}\lambda^2 \sum_{ab} \sigma_{ab}(t) \big[L_{ab}, \rho_S^0 \big] + \mathcal{O}\{\lambda^3\}, \tag{2.17}$$

where we have introduced the Lindblad jump operators $L_{ab} = |a\rangle\langle b|$ and the time-dependent coefficients

$$\gamma_{ab,cd}(t) = \sum_{\alpha\beta} \int_0^t dt_1 \int_0^t dt_2 \, C_{\alpha\beta}(t_1, t_2) \langle a | \boldsymbol{A}_\beta(t_2) | b \rangle \langle c | \boldsymbol{A}_\alpha^\dagger(t_1) | d \rangle^*,$$

$$\tag{2.18}$$

$$\sigma_{ab}(t) = \frac{-\mathrm{i}}{2} \sum_{\alpha\beta} \int_0^t dt_1 \int_0^t dt_2 \, C_{\alpha\beta}(t_1, t_2) \, \mathrm{sgn}(t_1 - t_2) \langle a | \boldsymbol{A}_\alpha(t_1) \boldsymbol{A}_\beta(t_2) | b \rangle,$$

which are evaluated most conveniently in the system energy eigenbasis. We note that we have neglected terms of higher than quadratic order λ, but the description remains consistent if we deliberately reinsert them, e.g., to preserve positivity.

We first show that the matrix $\gamma_{ab,cd}$ is positive semidefinite: this can be shown by $\sum_{ab,cd} x_{ab}^* \gamma_{ab,cd} x_{cd} = \sum_{A,B} x_A^* \gamma_{AB} x_B \geq 0$ (where we have introduced the short hand notation $A = (a,b)$ and $B = (c,d)$) for all times

$$
\sum_{ab,cd} x_{ab}^* \gamma_{ab,cd} x_{cd} = \sum_{\alpha\beta} \sum_{ab,cd} \int_0^t dt_1 \int_0^t dt_2 \, \text{Tr}_B \{ B_\alpha(t_1) B_\beta(t_2) \bar{\rho}_B \}
$$

$$
\times x_{ab}^* \langle a | A_\beta(t_2) | b \rangle x_{cd} \langle c | A_\alpha^\dagger(t_1) | d \rangle^*
$$

$$
= \text{Tr}_B \left\{ \left[\int_0^t dt_1 \sum_{cd} x_{cd} \langle d | \sum_\alpha A_\alpha(t_1) B_\alpha(t_1) | c \rangle \right] \right.
$$

$$
\left. \times \left[\int_0^t dt_2 \sum_{ab} x_{ab}^* \langle a | \sum_\beta A_\beta(t_2) B_\beta(t_2) | b \rangle \right] \bar{\rho}_B \right\}
$$

$$
= \text{Tr}_B \left\{ \left[\int_0^t dt_1 \sum_{cd} x_{cd} \langle d | \sum_\alpha A_\alpha(t_1) B_\alpha(t_1) | c \rangle \right] \right.
$$

$$
\left. \times \left[\int_0^t dt_2 \sum_{ab} x_{ab}^* \langle b | \sum_\beta A_\beta(t_2) B_\beta(t_2) | a \rangle^* \right] \bar{\rho}_B \right\}
$$

$$
= \text{Tr}_B \{ C^\dagger(t) C(t) \bar{\rho}_B \} \geq 0, \tag{2.19}
$$

where the last inequality holds for any operator $C(t)$ with a positive definite density matrix $\bar{\rho}_B$.

With the same short-hand notation, we can write the system density matrix as

$$
\rho_S(\Delta t) = \rho_S(0) + \lambda^2 \sum_{AB} \gamma_{AB}(\Delta t) \left[L_A \rho_S^0 L_B^\dagger - \frac{1}{2} \{ L_B^\dagger L_A, \rho_S^0 \} \right]
$$

$$
- i\lambda^2 [H_{\text{eff}}, \rho_S^0] + \mathcal{O}\{\lambda^3\}, \tag{2.20}
$$

where the introduced operator $H_{\text{eff}} = \sum_{ab} \sigma_{ab}(\Delta t) L_{ab}$ is hermitian. We define the operators

$$
K_1 = 1, \qquad K_2 = -\frac{1}{2} \sum_{AB} \gamma_{AB}(\Delta t) L_B^\dagger L_A - i H_{\text{eff}},
$$

$$
K_3 = L_1, \qquad K_{D+2} = L_D \tag{2.21}
$$

and rewrite the above equation as

$$
\rho_S(\Delta t) = \sum_{AB} w_{AB} K_A \rho_S^0 K_B^\dagger + \mathcal{O}\{\lambda^3\}, \tag{2.22}
$$

where the matrix

$$
w = \begin{pmatrix}
1 & \lambda^2 & 0 & \cdots & 0 \\
\lambda^2 & \lambda^4 & 0 & \cdots & 0 \\
0 & 0 & \lambda^2\gamma_{11} & \cdots & \lambda^2\gamma_{1D} \\
\vdots & \vdots & \vdots & \ddots & \vdots \\
0 & 0 & \lambda^2\gamma_{D1} & \cdots & \lambda^2\gamma_{DD}
\end{pmatrix}
\tag{2.23}
$$

is evidently positive definite: the larger lower right block is positive semidefinite due to the positivity of γ_{AB}. The upper left block has eigenvalues $\lambda_0 = 0$ and $\lambda_1 = 1 + \lambda^4$, as can be easily confirmed. If we had—as is usual for perturbation— neglected the w_{22} element completely, one of its eigenvalues would remain negative. In principle, we could have inserted larger values for w_{22} as long as they were not larger than $\mathcal{O}\{\lambda^3\}$ while remaining consistent with our truncation order. However, here we aim at preserving positivity while applying the smallest possible modifi- cation to our equations, and under this side constraint the above choice becomes unique. Unfortunately, we note that due to this additional term the trace is no longer conserved:

$$
\sum_{\alpha\beta} w_{\alpha\beta} K_\beta^\dagger K_\alpha = 1 + \lambda^4 K_2^\dagger K_2.
\tag{2.24}
$$

Though this correction is small, repeated application of the map defined this way would result in an exploding trace of the density matrix, such that the probability interpretation would no longer be valid. We take a closer look at the operator K_2 by inserting all necessary definitions,

$$
\begin{aligned}
K_2 &= \frac{-1}{2}\sum_{ab,cd}\gamma_{ab,cd}(\Delta t)L_{cd}^\dagger L_{ab} - i\sum_{ab}\sigma_{ab}(\Delta t)L_{ab} \\
&= \sum_{ab}\left(\frac{-1}{2}\sum_c \gamma_{cb,ca}(\Delta t) - i\sigma_{ab}(\Delta t)\right)L_{ab} \\
&= -\sum_{\alpha\beta}\int_0^{\Delta t}dt_1 \int_0^{\Delta t}dt_2\, C_{\alpha\beta}(t_1,t_2)\Theta(t_1 - t_2)A_\alpha(t_1)A_\beta(t_2).
\end{aligned}
\tag{2.25}
$$

To obtain a map that preserves the probability interpretation of the density ma- trix, we therefore have to renormalize the density matrix after each application. In addition, we assume that the bath part is roughly unaffected by the evolution, such that we may use the map

$$
\rho_S(t + \Delta t) = \frac{\sum_{\alpha\beta} w_{\alpha\beta} K_\alpha \rho_S(t) K_\beta^\dagger}{\mathrm{Tr}\{\sum_{\alpha\beta} w_{\alpha\beta} K_\alpha \rho_S(t) K_\beta^\dagger\}}
\tag{2.26}
$$

for all times t. This will also preserve the trace of the density matrix, at the price of obtaining a nonlinear map. However, we note that in the continuum limit, where

$\Delta t \to 0$, the non-preservation of the trace will become negligible, and renormalization may not be necessary.

Now, putting it all together (i.e., undoing the scaling transformation $\lambda \to 1$ and inserting all definitions), we have generated a fixed-point iteration scheme for the density matrix that will always preserve its positivity, trace, and hermiticity.

Definition 2.1 (Map for short times/weak couplings) In the weak coupling limit with a decomposition of the interaction Hamiltonian $H_{\rm I} = \sum_\alpha A_\alpha \otimes B_\alpha$, the density matrix in the interaction picture obeys the map

$$\rho_{\rm S}(t + \Delta t) = \frac{\sum_{\alpha\beta} w_{\alpha\beta} K_\alpha \rho_{\rm S} K_\beta^\dagger}{{\rm Tr}\{\sum_{\alpha\beta} w_{\alpha\beta} K_\alpha \rho_{\rm S} K_\beta^\dagger\}}, \tag{2.27}$$

where

$$w = \begin{pmatrix} 1 & 1 & 0 & \cdots & 0 \\ 1 & 1 & 0 & \cdots & 0 \\ \hline 0 & 0 & \gamma_{11,11} & \cdots & \gamma_{11,DD} \\ \vdots & \vdots & \vdots & \ddots & \vdots \\ 0 & 0 & \gamma_{DD,11} & \cdots & \gamma_{DD,DD} \end{pmatrix},$$

$$\gamma_{ab,cd}(t) = \sum_{\alpha\beta} \int_0^t dt_1 \int_0^t dt_2\, C_{\alpha\beta}(t_1, t_2)\langle a|A_\beta(t_2)|b\rangle\langle c|A_\alpha^\dagger(t_1)|d\rangle^*,$$

$$\sigma_{ab}(t) = \frac{-{\rm i}}{2} \sum_{\alpha\beta} \int_0^t dt_1 \int_0^t dt_2\, C_{\alpha\beta}(t_1, t_2)\,{\rm sgn}(t_1 - t_2)\langle a|A_\alpha(t_1)A_\beta(t_2)|b\rangle, \tag{2.28}$$

$$K_1 = \mathbf{1},$$

$$K_2 = -\sum_{\alpha\beta} \int_0^{\Delta t} dt_1 \int_0^{\Delta t} dt_2\, C_{\alpha\beta}(t_1, t_2)\Theta(t_1 - t_2)A_\alpha(t_1)A_\beta(t_2),$$

$$K_3 = |1\rangle\langle 1|, \qquad \ldots, \qquad K_{2+D} = |1\rangle\langle D|,$$

$$K_{3+D} = |2\rangle\langle 1|, \qquad \ldots, \qquad K_{2+2D} = |2\rangle\langle D|,$$

$$\vdots$$

$$K_{3+(D-1)D} = |D\rangle\langle 1|, \qquad \ldots, \qquad K_{2+D^2} = |D\rangle\langle D|,$$

where $|1\rangle, \ldots, |D\rangle$ represents an orthonormal basis in the D-dimensional system Hilbert space.

We note that the matrix elements in the above definition are most conveniently evaluated in the energy eigenbasis of the system $H_{\rm S}|n\rangle = E_n|n\rangle$.

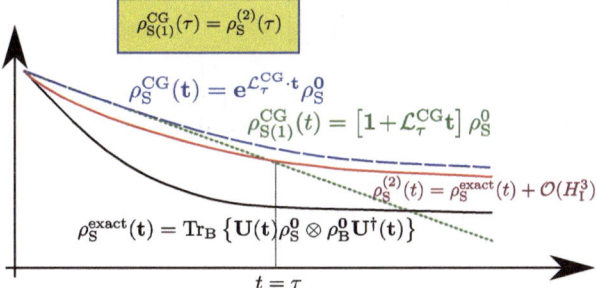

Fig. 2.1 Sketch of the coarse-graining approximation scheme. Calculating the exact time evolution operator in Eq. (1.5) in a closed form is usually prohibitive, which renders the calculation of the exact solution (*solid black curve*) an impossible task. It is however possible to expand the evolution operator $U(t) = \mathbf{1} - i\int_0^t \boldsymbol{H}_I(t')\,dt' - \int_0^t dt_1\,dt_2\,\boldsymbol{H}_I(t_1)\boldsymbol{H}_I(t_2)\Theta(t_1 - t_2) + \mathcal{O}\{H_I^3\}$ to second order in the interaction $\boldsymbol{H}_I$ and to obtain the corresponding reduced approximate density matrix (*solid red curve*). Calculating the matrix exponential of a constant Lindblad-type generator $\mathcal{L}_\tau^{CG}$ is also usually prohibitive (*dashed blue curve*), but the first-order approximation (*dotted green line*) may be matched with the weak coupling approximation of the exact solution at time $t = \tau$ to obtain a defining equation for $\mathcal{L}_\tau^{CG}$

2.3 Master Equation in the Weak Coupling Limit

The standard derivation in the weak coupling limit consists of applying the Born, Markov, and secular approximations in a certain sequence, which eventually yields a Lindblad master equation with further appealing properties [2]. Here, we will present an alternative scheme based on coarse graining [3]. As an adjustable parameter, this involves a coarse-graining time scale τ, after which the reduced density matrix should be closest to the exact solution. It can be shown [4] that the limit $\tau \to \infty$ reproduces the standard textbook results. For finite coarse-graining times however one obtains Lindblad form master equations that yield valid short-time descriptions [4, 5]. See Fig. 2.1.

2.3.1 Coarse-Graining Master Equation

We use the result derived in the previous section, Eq. (2.16). At time t, this should for weak coupling match the evolution by a Markovian generator,

$$\rho_S^{CG}(\tau) = e^{\mathcal{L}_\tau^{CG}\cdot\tau}\rho_S^0 \approx \left[\mathbf{1} + \mathcal{L}_\tau^{CG}\cdot\tau\right]\rho_S^0, \qquad (2.29)$$

such that we can infer the action of the generator on an arbitrary density matrix

$$\mathcal{L}_\tau^{CG}\boldsymbol{\rho}_S = \frac{1}{\tau}\sum_{\alpha\beta}\int_0^\tau dt_1\int_0^\tau dt_2\, C_{\alpha\beta}(t_1,t_2)\big[A_\beta(t_2)\boldsymbol{\rho}_S A_\alpha(t_1)$$

$$-\Theta(t_1 - t_2)A_\alpha(t_1)A_\beta(t_2)\boldsymbol{\rho}_S - \Theta(t_2 - t_1)\boldsymbol{\rho}_S A_\alpha(t_1)A_\beta(t_2)\big]$$

$$= -i\left[\frac{1}{2i\tau}\int_0^\tau dt_1 \int_0^\tau dt_2 \, \text{sgn}(t_1 - t_2) \sum_{\alpha\beta} C_{\alpha\beta}(t_1, t_2) A_\alpha(t_1) A_\beta(t_2), \rho_S\right]$$

$$+ \frac{1}{\tau}\int_0^\tau dt_1 \int_0^\tau dt_2 \sum_{\alpha\beta} C_{\alpha\beta}(t_1, t_2)\left[A_\beta(t_2)\rho_S A_\alpha(t_1)\right.$$

$$\left.- \frac{1}{2}\left\{A_\alpha(t_1)A_\beta(t_2), \rho_S\right\}\right], \tag{2.30}$$

where we have inserted $\Theta(x) = \frac{1}{2}[1 + \text{sgn}(x)]$. The above equation can be taken as the definition of a coarse-graining Lindblad generator. To show its Lindblad form, we first note that the effective Hamiltonian

$$H_{\text{eff}}^\tau = \frac{1}{2i\tau}\int_0^\tau dt_1 \int_0^\tau dt_2 \, \text{sgn}(t_1 - t_2) \sum_{\alpha\beta} C_{\alpha\beta}(t_1, t_2) A_\alpha(t_1) A_\beta(t_2) \tag{2.31}$$

in the commutator is hermitian, which can be seen by reinserting the definitions of the two-argument correlation functions (2.15). To show that the second line can be written as a Lindblad dissipator, we insert identities in a time-independent basis:

$$A_\alpha(t_1) = \sum_{cd} |d\rangle\langle d|A_\alpha(t_1)|c\rangle\langle c|,$$

$$A_\beta(t_2) = \sum_{ab} |a\rangle\langle a|A_\beta(t_2)|b\rangle\langle b|, \tag{2.32}$$

such that the coarse-graining Liouvillian can be written as

$$\mathscr{L}_\tau^{\text{CG}} \rho_S = -i\left[H_{\text{eff}}^\tau, \rho_S\right] + \sum_{abcd} \gamma_{ab,cd}^\tau \left[L_{ab}\rho_S L_{cd}^\dagger - \frac{1}{2}\left\{L_{cd}^\dagger L_{ab}, \rho_S\right\}\right] \tag{2.33}$$

with Lindblad jumpers $L_{ab} = |a\rangle\langle b|$ and the dissipation coefficients

$$\gamma_{ab,cd}^\tau = \frac{1}{\tau}\int_0^\tau dt_1 \int_0^\tau dt_2 \sum_{\alpha\beta} C_{\alpha\beta}(t_1, t_2)\langle a|A_\beta(t_2)|b\rangle\langle d|A_\alpha(t_1)|c\rangle. \tag{2.34}$$

To prove the Lindblad form we simply have to demonstrate positivity of the damping matrix:

$$\sum_{ab,cd} x_{ab}^* \gamma_{ab,cd} x_{cd} = \text{Tr}_B\left\{\left(\int_0^\tau dt_1 \sum_{cd}\sum_\alpha B_\alpha(t_1)x_{cd}\langle d|A_\alpha(t_1)|c\rangle\right)\right.$$

$$\left.\times\left(\int_0^\tau dt_2 \sum_{ab}\sum_\beta B_\beta(t_2)x_{ab}^*\langle a|A_\beta(t_2)|b\rangle\right)\bar\rho_B\right\}$$

$$= \text{Tr}_B \left\{ \left(\int_0^\tau dt_1 \sum_{cd} x_{cd} \langle d | \boldsymbol{H}_I(t_1) | c \rangle \right) \right.$$

$$\left. \times \left(\int_0^\tau dt_2 \sum_{ab} x_{ab}^* \langle a | \boldsymbol{H}_I(t_2) | b \rangle \right) \bar{\rho}_B \right\}$$

$$= \text{Tr} \left\{ B^\dagger(\tau) B(\tau) \bar{\rho}_B \right\} \geq 0, \tag{2.35}$$

where we have introduced the auxiliary operator $B(\tau) = \int_0^\tau dt_2 \sum_{ab} x_{ab}^* \times \langle a | \boldsymbol{H}_I(t_2) | b \rangle$ and where the last line follows from positivity of the reservoir density matrix. Therefore, we conclude that coarse graining always yields a Lindblad-type master equation, regardless of the bath state, hermiticity of coupling operators, etc.

For most applications however, the form of Eq. (2.30) is not quite practical, since for example we have not yet used the fact that the bath correlation functions typically only depend on a single argument. One important situation in which this is always the case is that of a single or multiple thermal reservoirs.

Exercise 2.3 (Properties of correlation functions) Show that when $[H_B, \bar{\rho}_B] = 0$ (which is, e.g., the case in thermal equilibrium), the correlation functions in Eq. (2.15) only depend on the difference of their time arguments

$$C_{\alpha\beta}(t_1, t_2) = C_{\alpha\beta}(t_1 - t_2, 0). \tag{2.36}$$

However, for a system that is in a non-equilibrium steady state one may also observe that in the long-time limit, the correlation functions only depend on the difference of their time arguments (see the discussion in Sect. 5.6).

For the case where the correlation functions only depend on the difference of their time arguments, $C_{\alpha\beta}(t_1 - t_2) \equiv C_{\alpha\beta}(t_1 - t_2, 0) = C_{\alpha\beta}(t_1, t_2)$, we introduce the even and odd Fourier transforms of the correlation functions, respectively, as

$$\gamma_{\alpha\beta}(\omega) = \int C_{\alpha\beta}(\tau) e^{+i\omega\tau} \, d\tau, \qquad \sigma_{\alpha\beta}(\omega) = \int C_{\alpha\beta}(\tau) \, \text{sgn}(\tau) e^{+i\omega\tau} \, d\tau. \tag{2.37}$$

We note here for later reference that the odd Fourier transform may be obtained from the even one:

$$\sigma_{\alpha\beta}(\omega) = \frac{i}{\pi} \mathscr{P} \int_{-\infty}^{+\infty} \frac{\gamma_{\alpha\beta}(\Omega)}{\omega - \Omega} \, d\Omega, \tag{2.38}$$

where $\mathscr{P}$ denotes the Cauchy principal value of the integral. Furthermore, we use the system energy eigenbasis $H_S | a \rangle = E_a | a \rangle$ to explicitly perform the temporal integrations. The Liouvillian then acts like

$$\dot{\boldsymbol{\rho}}_S = -i \frac{1}{4i\pi\tau} \int d\omega \sum_{abc} \int_0^\tau dt_1 \int_0^\tau dt_2 \sum_{\alpha\beta} \sigma_{\alpha\beta}(\omega)$$

$$\times e^{-i\omega(t_1 - t_2)} e^{+i(E_a - E_c)t_1} e^{+i(E_c - E_b)t_2} \langle c | A_\beta | b \rangle \langle c | A_\alpha^\dagger | a \rangle^* \left[| a \rangle \langle b |, \boldsymbol{\rho}_S \right]$$

$$+ \frac{1}{2\pi\tau} \int d\omega \int_0^\tau dt_1 \int_0^\tau dt_2$$

$$\times \sum_{\alpha\beta} \sum_{abcd} \gamma_{\alpha\beta}(\omega) e^{-i\omega(t_1-t_2)} e^{+i(E_a-E_b)t_2} e^{+i(E_d-E_c)t_1} \langle a|A_\beta|b\rangle \langle c|A_\alpha^\dagger|d\rangle^*$$

$$\times \left[|a\rangle\langle b|\rho_S(|c\rangle\langle d|)^\dagger - \frac{1}{2}\{(|c\rangle\langle d|)^\dagger|a\rangle\langle b|, \rho_S\} \right]. \tag{2.39}$$

We perform the temporal integrations by invoking

$$\int_0^\tau e^{i\alpha_k t_k} dt_k = \tau e^{i\alpha_k\tau/2} \operatorname{sinc}\left[\frac{\alpha_k\tau}{2}\right] \tag{2.40}$$

with the band-filter function $\operatorname{sinc}(x) \equiv \sin(x)/x$ to obtain

$$\dot{\rho}_S = -i\frac{\tau}{4i\pi} \int d\omega \sum_{abc} \sum_{\alpha\beta} \sigma_{\alpha\beta}(\omega) e^{i\tau(E_a-E_b)/2}$$

$$\times \operatorname{sinc}\left[\frac{\tau}{2}(E_a - E_c - \omega)\right] \operatorname{sinc}\left[\frac{\tau}{2}(E_c - E_b + \omega)\right]$$

$$\times \langle c|A_\beta|b\rangle \langle c|A_\alpha^\dagger|a\rangle^* \left[|a\rangle\langle b|, \rho_S\right]$$

$$+ \frac{\tau}{2\pi} \int d\omega \sum_{\alpha\beta} \sum_{abcd} \gamma_{\alpha\beta}(\omega) e^{i\tau(E_a-E_b+E_d-E_c)/2}$$

$$\times \operatorname{sinc}\left[\frac{\tau}{2}(E_d - E_c - \omega)\right] \operatorname{sinc}\left[\frac{\tau}{2}(\omega + E_a - E_b)\right]$$

$$\times \langle a|A_\beta|b\rangle \langle c|A_\alpha^\dagger|d\rangle^* \left[|a\rangle\langle b|\rho_S(|c\rangle\langle d|)^\dagger - \frac{1}{2}\{(|c\rangle\langle d|)^\dagger|a\rangle\langle b|, \rho_S\}\right]. \tag{2.41}$$

This expression is already quite similar to the standard quantum optical master equation [4], but here the dampening coefficients are expressed in terms of integrals that depend on the coarse-graining time scale τ.

Definition 2.2 (Coarse-graining Liouvillian) For a system-bath interaction of the form $H_I = \sum_\alpha A_\alpha \otimes B_\alpha$ and a stationary reservoir density matrix obeying $[H_B, \bar{\rho}_B] = 0$ and $\operatorname{Tr}\{B_\alpha\bar{\rho}_B\} = 0$, the coarse-graining Lindblad Liouvillian is in the system energy eigenbasis $H_S|a\rangle = E_a|a\rangle$ given by

$$\dot{\rho}_S = -i\left[\sum_{ab} \sigma_{ab}^\tau |a\rangle\langle b|, \rho_S\right]$$

$$+ \sum_{abcd} \gamma_{ab,cd}^\tau \left[|a\rangle\langle b|\rho_S(|c\rangle\langle d|)^\dagger - \frac{1}{2}\{(|c\rangle\langle d|)^\dagger|a\rangle\langle b|, \rho_S\}\right] \tag{2.42}$$

with the coefficients

$$
\sigma_{ab}^{\tau} = \frac{1}{2i} \int d\omega \sum_{c} e^{i\tau(E_a - E_b)/2} \frac{\tau}{2\pi} \mathrm{sinc}\left[\frac{\tau}{2}(E_a - E_c - \omega)\right]
$$

$$
\times \mathrm{sinc}\left[\frac{\tau}{2}(E_b - E_c - \omega)\right] \left[\sum_{\alpha\beta} \sigma_{\alpha\beta}(\omega)\langle c|A_\beta|b\rangle\langle c|A_\alpha^\dagger|a\rangle^*\right],
$$

$$
\gamma_{ab,cd}^{\tau} = \int d\omega\, e^{i\tau(E_a - E_b + E_d - E_c)/2} \frac{\tau}{2\pi} \mathrm{sinc}\left[\frac{\tau}{2}(E_d - E_c - \omega)\right]
$$

$$
\times \mathrm{sinc}\left[\frac{\tau}{2}(E_b - E_a - \omega)\right] \left[\sum_{\alpha\beta} \gamma_{\alpha\beta}(\omega)\langle a|A_\beta|b\rangle\langle c|A_\alpha^\dagger|d\rangle^*\right].
$$

(2.43)

When transforming the master equation back to the Schrödinger picture $\rho_S(t) = e^{-iH_S t}\rho_S(t)e^{+iH_S t}$, we will get the system Hamiltonian in the commutator, too, and some additional phases that lead to time-dependent coefficients but do not destroy the Lindblad form. We note that this Lindblad master equation will in general not assume a rate equation form in the system energy eigenbasis. The theory contains the coarse-graining time τ as a free parameter, and a sufficient condition for the validity of the perturbative treatment is $\|\mathscr{L}_\tau^{CG}\tau\| \ll 1$, but it is not a necessary condition.

2.3.2 Quantum Optical Master Equation

When the coarse-graining time is sent to infinity, $\tau \to \infty$, one can show that the standard quantum optical master equation (that normally arises when Born, Markov, and secular approximations are applied) is recovered: in our case, this simply requires us to use the identity

$$
\lim_{\tau \to \infty} \tau\, \mathrm{sinc}\left[\frac{\tau}{2}(\Omega_a - \omega)\right] \mathrm{sinc}\left[\frac{\tau}{2}(\Omega_b - \omega)\right] = 2\pi \delta_{\Omega_a,\Omega_b}\delta(\Omega_a - \omega)   \quad (2.44)
$$

to collapse the integrals in (2.43). It is also visible that, when transforming back to the Schrödinger picture, the time-dependent phases cancel due to the arising Kronecker functions.

Definition 2.3 (Quantum optical master equation)   In the weak coupling limit, an interaction Hamiltonian of the form $H_I = \sum_\alpha A_\alpha \otimes B_\alpha$ obeying $[H_B, \bar{\rho}_B] = 0$ and $\mathrm{Tr}\{B_\alpha \bar{\rho}_B\} = 0$ leads in the system energy eigenbasis $H_S|a\rangle = E_a|a\rangle$ to the Lindblad form master equation in the Schrödinger picture:

$$\dot{\rho}_S = -i\left[H_S + \sum_{ab}\sigma_{ab}|a\rangle\langle b|, \rho_S(t)\right]$$

$$+ \sum_{a,b,c,d}\gamma_{ab,cd}\left[|a\rangle\langle b|\rho_S(t)\left(|c\rangle\langle d|\right)^\dagger - \frac{1}{2}\left\{\left(|c\rangle\langle d|\right)^\dagger|a\rangle\langle b|, \rho_S(t)\right\}\right],$$

$$\tag{2.45}$$

$$\gamma_{ab,cd} = \sum_{\alpha\beta}\gamma_{\alpha\beta}(E_b - E_a)\delta_{E_b-E_a,E_d-E_c}\langle a|A_\beta|b\rangle\langle c|A_\alpha^\dagger|d\rangle^*,$$

$$\sigma_{ab} = \sum_{\alpha\beta}\sum_{c}\frac{1}{2i}\sigma_{\alpha\beta}(E_b - E_c)\delta_{E_b,E_a}\langle c|A_\beta|b\rangle\langle c|A_\alpha|a\rangle^*,$$

where $H_{LS} = \sum_{ab}\sigma_{ab}|a\rangle\langle b| = H_{LS}^\dagger$ is also called the Lamb-shift Hamiltonian.

2.3.3 Properties of the Quantum Optical Master Equation

2.3.3.1 Pointer Basis

The simple master equation in Definition 2.3 is not only very popular due to its simple computability. First, we note that the Lamb-shift Hamiltonian commutes with the system Hamiltonian $[H_S, H_{LS}] = 0$. This implies that there exists a basis diagonalizing both operators.

In particular, in the case when the system is already nondegenerate, the basis diagonalizing H_S is unique and thus also diagonalizes H_{LS}. By using $\delta_{E_a,E_b} \to \delta_{a,b}$ and $\delta_{E_a-E_b,E_a-E_c} = \delta_{E_b,E_c} \to \delta_{b,c}$, we see that, in this case, only the diagonals of the density matrix in the energy eigenbasis couple to themselves:

$$\dot{\rho}_{aa} = \sum_{b}\gamma_{ab,ab}\rho_{bb} - \left[\sum_{b}\gamma_{ba,ba}\right]\rho_{aa}, \tag{2.46}$$

whereas the coherences usually simply decay. Such rate equations are of course much simpler to solve than the full master equation.

Definition 2.4 (Quantum optical rate equation) In the weak coupling limit, an interaction Hamiltonian of the form $H_I = \sum_\alpha A_\alpha \otimes B_\alpha$ obeying $[H_B, \bar{\rho}_B] = 0$ and $\text{Tr}\{B_\alpha\bar{\rho}_B\} = 0$ leads for a nondegenerate system in the system energy eigenbasis $H_S|a\rangle = E_a|a\rangle$ to a rate equation for the populations of the density matrix

$$\dot{\rho}_{aa} = \sum_{b}\gamma_{ab,ab}\rho_{bb} - \left[\sum_{b}\gamma_{ba,ba}\right]\rho_{aa}, \tag{2.47}$$

where the rates are given by matrix elements of the coupling operators and the even Fourier transform of the correlation function (2.37),

$$\gamma_{ab,ab} = \sum_{\alpha\beta} \gamma_{\alpha\beta}(E_b - E_a)\langle a|A_\beta|b\rangle\langle a|A_\alpha^\dagger|b\rangle^* \geq 0. \tag{2.48}$$

In a basis different from the system energy eigenbasis, the simple rate equation picture does not apply.

We note that the positivity of the transition rates follows from the fact that $\gamma_{ab,ab}$ are just the diagonal entries of the damping matrix $\gamma_{ab,cd}$, for which we have already demonstrated positivity: diagonal entries of non-negative matrices are also non-negative. This rate equation representation implies that for the assumed prerequisites (weak coupling, no degeneracies), the system energy eigenbasis quite generally becomes the pointer basis (i.e., the basis within which the stationary density matrix becomes diagonal). With degeneracies present in H_S, the type of interaction may determine the pointer basis for extremely small coupling strengths [5, 6].

Furthermore, we note that a system described by a rate equation may still exhibit complex quantum behavior. This can for example be expected when the basis within which this rate equation applies exhibits true quantum properties such as entanglement [7, 8].

2.3.3.2 Steady State for a Thermal Reservoir

Furthermore, for a single reservoir in thermal equilibrium

$$\bar{\rho}_B = \frac{e^{-\beta H_B}}{\mathrm{Tr}\{e^{-\beta H_B}\}} \tag{2.49}$$

a stationary state of the rate equation (2.47) is the thermal one

$$\bar{\rho}_S = \frac{e^{-\beta H_S}}{\mathrm{Tr}\{e^{-\beta H_S}\}} \tag{2.50}$$

with exactly the same temperature as that of the reservoir.

Formally, this can be traced back to analytic properties of the bath correlation functions. For thermal reservoirs, these obey Kubo–Martin–Schwinger (KMS) conditions

$$C_{\alpha\beta}(\tau) = C_{\beta\alpha}(-\tau - i\beta). \tag{2.51}$$

Exercise 2.4 (KMS condition) Show the validity of the KMS condition for a thermal bath with $\bar{\rho}_B = \frac{e^{-\beta H_B}}{\mathrm{Tr}\{e^{-\beta H_B}\}}$.

This shift property implies the following for the Fourier transforms of the bath correlation function:

$$\gamma_{\alpha\beta}(-\omega) = \gamma_{\beta\alpha}(+\omega)e^{-\beta\omega}, \tag{2.52}$$

such that eventually the transition rates—compare Eq. (2.45)—from b to a in the rate equation

$$\gamma_{ab,ab} = \sum_{\alpha\beta} \gamma_{\alpha\beta}(E_b - E_a)\langle a|A_\beta|b\rangle\langle a|A_\alpha^\dagger|b\rangle^* \tag{2.53}$$

obey global detailed balance relations

$$\frac{\gamma_{ab,ab}}{\gamma_{ba,ba}} = e^{\beta(E_b - E_a)}, \tag{2.54}$$

which can be used to prove equilibration of the system temperature with that of the bath.

Furthermore, when the bath is in a grand-canonical ensemble,

$$\bar{\rho}_B = \frac{e^{-\beta(H_B - \mu N_B)}}{\mathrm{Tr}\{e^{-\beta(H_B - \mu N_B)}\}}, \tag{2.55}$$

where β denotes the inverse temperature, μ the chemical potential, and N_B the particle number of the bath, one can show [9]—given that the total particle number is conserved: $[H_S, N_S] = 0$, $[H_B, N_B] = 0$, and $[H_I, N_S + N_B] = 0$—that a stationary state is also present when both temperature and chemical potential are equilibrated:

$$\bar{\rho}_S = \frac{e^{-\beta(H_S - \mu N_S)}}{\mathrm{Tr}\{e^{-\beta(H_S - \mu N_S)}\}}. \tag{2.56}$$

In general, this does not exclude the existence of further stationary states, but with a sufficiently complex coupling between system and reservoir, one can expect the above stationary solution to be unique [2].

2.3.3.3 Multiple Thermal Reservoirs

When a system is coupled to multiple reservoirs that are held at different equilibrium states,

$$\bar{\rho}_B = \bar{\rho}_B^{(1)} \otimes \cdots \otimes \bar{\rho}_B^{(N)}, \tag{2.57}$$

all the previous derivations go through, but we note here that we can enforce the previously mentioned constraint $\langle B_\alpha^{(\nu)}\rangle = 0$ for all bath operators $B_\alpha^{(\nu)}$ and all reservoirs ν. This implies that the correlation function involving bath operators acting on different reservoirs will vanish,

$$C_{\alpha\beta}^{(\mu\nu)}(\tau) = \langle B_\alpha^{(\mu)}(\tau)B_\beta^{(\nu)}\rangle = 0, \quad \text{when } \mu \neq \nu. \tag{2.58}$$

Keeping in mind that the Fourier transforms of these quantities are linear, this implies that within the already-used limitations the resulting Liouvillian can be represented as an additive combination of the separate dissipators,

$$\mathscr{L} = \sum_{\nu} \mathscr{L}^{(\nu)}, \tag{2.59}$$

where $\mathscr{L}^{(\nu)}$ is the dissipator if the system was only coupled to the single reservoir ν. If for example the system has all the favorable properties enabling the derivation of a rate equation, this decomposition of course also transfers to the rates:

$$\gamma_{ab,ab} = \sum_{\nu} \gamma_{ab,ab}^{(\nu)}. \tag{2.60}$$

The interesting consequence of this procedure is now that when the reservoirs are held at different equilibrium states, the system experiences a highly non-equilibrium environment, which induces, e.g., steady-state energy and matter currents. In this case, the separate detailed balance relations will still hold:

$$\frac{\gamma_{ab,ab}^{(\nu)}}{\gamma_{ba,ba}^{(\nu)}} = e^{\beta^{\nu}(E_b - E_a)}, \tag{2.61}$$

but of course the global detailed balance relation will in general be violated:

$$\frac{\gamma_{ab,ab}}{\gamma_{ba,ba}} \neq e^{\beta(E_b - E_a)}. \tag{2.62}$$

There exist a few special cases however (e.g., systems described by only a single transition frequency), where one obtains thermalization at some average temperature [7].

2.4 Strong Coupling Limit

In the previous derivation, we have used the fact that the interaction Hamiltonian is weak in comparison to both the system part and the reservoir part. The latter assumption is required to keep the reservoir at a stationary limit, whereas the first assumption may be inverted. It is for example still possible to derive a master equation in the limit where the interaction dominates the system Hamiltonian. Within the framework of coarse graining, we can apply exactly the same formalism, but should now keep in mind that oscillations of the system are much slower than the ones induced by the interaction. This implies that we can neglect the time dependence by inserting $A_{\alpha}(t) \to A_{\alpha}$, which corresponds to setting the system energies to zero in Eq. (2.39). The Liouvillian in the interaction picture then becomes

$$\dot{\rho}_S = -i\frac{\tau}{4\pi i} \int \sum_{\alpha\beta} \sigma_{\alpha\beta}(\omega) \operatorname{sinc}^2\left[\frac{\omega\tau}{2}\right] d\omega [A_\alpha A_\beta, \rho_S]$$

$$+ \frac{\tau}{2\pi} \int \sum_{\alpha\beta} \gamma_{\alpha\beta}(\omega) \operatorname{sinc}^2\left[\frac{\omega\tau}{2}\right] d\omega \left[A_\beta \rho_S A_\alpha - \frac{1}{2}\{A_\alpha A_\beta, \rho_S\}\right].$$

(2.63)

Now inserting the limit of large coarse-graining times,

$$\lim_{\tau\to\infty} \tau \operatorname{sinc}^2\left[\frac{\omega\tau}{2}\right] = 2\pi\delta(\omega),$$

(2.64)

the Liouvillian in the strong coupling limit can be readily calculated:

$$\mathscr{L}\rho_S = -i\sum_{\alpha\beta} \frac{\sigma_{\alpha\beta}(0)}{2i}[A_\alpha A_\beta, \rho_S]$$

$$+ \sum_{\alpha\beta} \gamma_{\alpha\beta}(0)\left[A_\beta \rho_S A_\alpha - \frac{1}{2}\{A_\alpha A_\beta, \rho_S\}\right].$$

(2.65)

Finally, transforming back to the Schrödinger picture (again neglecting the small system eigenenergies), one obtains a simple form for the generator in the strong coupling limit. Alternatively, the generator can be obtained by a scaling transformation, from which it is also often called the singular coupling limit.

Definition 2.5 (Master equation in the singular coupling limit) In the singular coupling limit, an interaction Hamiltonian of the form $H_I = \sum_\alpha A_\alpha \otimes B_\alpha$ yields the Liouvillian in the Schrödinger picture

$$\mathscr{L}\rho_S = -i[H_S, \rho_S] - i\sum_{\alpha\beta} \frac{\sigma_{\alpha\beta}(0)}{2i}[A_\alpha A_\beta, \rho_S]$$

$$+ \sum_{\alpha\beta} \gamma_{\alpha\beta}(0)\left[A_\beta \rho_S A_\alpha - \frac{1}{2}\{A_\alpha A_\beta, \rho_S\}\right].$$

References

1. E.B. Davies, Markovian master equations. Commun. Math. Phys. **39**, 91 (1974)
2. H.-P. Breuer, F. Petruccione, *The Theory of Open Quantum Systems* (Oxford University Press, Oxford, 2002)
3. D.A. Lidar, Z. Bihary, K.B. Whaley, From completely positive maps to the quantum Markovian semigroup master equation. Chem. Phys. **268**, 35 (2001)
4. G. Schaller, T. Brandes, Preservation of positivity by dynamical coarse-graining. Phys. Rev. A **78**, 022106 (2008)
5. G. Schaller, P. Zedler, T. Brandes, Systematic perturbation theory for dynamical coarse-graining. Phys. Rev. A **79**, 032110 (2009)

6. M.G. Schultz, F. von Oppen, Quantum transport through nanostructures in the singular-coupling limit. Phys. Rev. B **80**, 033302 (2009)
7. M. Vogl, G. Schaller, T. Brandes, Counting statistics of collective photon transmissions. Ann. Phys. **326**, 2827 (2011)
8. M. Vogl, G. Schaller, T. Brandes, Criticality in transport through the quantum Ising chain. Phys. Rev. Lett. **109**, 240402 (2012)
9. G. Schaller, Quantum equilibration under constraints and transport balance. Phys. Rev. E **83**, 031111 (2011)

References

Chapter 3
Exactly Solvable Models

Abstract To understand the limit within which master equations are valid, it is quite instructive to compare the master equation results against exactly solvable models. Unfortunately, these models are quite rare. In this chapter, we will discuss two popular representatives of exactly solvable models: first, we investigate a pure dephasing spin-boson model, where the interaction Hamiltonian commutes with the system Hamiltonian. Such models obviously leave the system energy invariant but nevertheless may be used to investigate interesting features such as decoherence. Second, we consider a noninteracting model, where the Hamiltonian can be written as a quadratic form of fermionic annihilation and creation operators. Such models generally admit—at least formally—an exact solution, and can thus be used to study non-equilibrium setups and transport in a regime where the coupling between system and reservoir becomes strong. Furthermore, we note that the non-equilibrium stationary solution of these models may also define a non-equilibrium reservoir.

3.1 Pure Dephasing Spin-Boson Model

The pure dephasing spin-boson model describes the interaction of a two-level system with a bosonic bath:

$$H_S = \omega \sigma^z,$$

$$H_B = \sum_k \omega_k \left(b_k^\dagger b_k + 1/2 \right),$$

$$H_I = \sigma^z \otimes \sum_k \left(h_k b_k + h_k^* b_k^\dagger \right),$$

(3.1)

where σ^z is a Pauli matrix and b_k a bosonic annihilation operator in the bath. One immediately observes that the model conserves the system energy—since $[H_S, H_I] = 0$—and will thus only modify the evolution of coherences in the system energy eigenbasis (hence the name purely dephasing). Similar models have been used to illustrate decoherence in quantum computers [1, 2] or to test the validity of Markovian master equations [3].

G. Schaller, *Open Quantum Systems Far from Equilibrium*, Lecture Notes in Physics 881,   47
DOI 10.1007/978-3-319-03877-3_3,
© Springer International Publishing Switzerland 2014

3.1.1 Time Evolution Operator

The calculation of the exact solution makes use of the fact that in the interaction picture, the time evolution operator can be exactly determined. In the interaction picture, the full density matrix follows the von Neumann equation

$$\dot{\rho} = -i\big[\boldsymbol{H}_I(t), \rho(t)\big] \tag{3.2}$$

with the interaction Hamiltonian in the interaction picture

$$\boldsymbol{H}_I(t) = \sigma^z \otimes \sum_k \big(h_k b_k e^{-i\omega_k \tau} + h_k^* b_k^\dagger e^{+i\omega_k \tau}\big). \tag{3.3}$$

Exercise 3.1 (Interaction picture) Show that Eq. (3.3) arises in the interaction picture.

We note that the commutator of the interaction Hamiltonian with itself at different times is just a number,

$$\big[\boldsymbol{H}_I(t_1), \boldsymbol{H}_I(t_2)\big] = \sum_k |h_k|^2 2i \sin\big[\omega_k(t_2 - t_1)\big], \tag{3.4}$$

such that the Baker–Campbell–Hausdorff (BCH) formula may be employed to calculate the exponential. For two operators A and B with the commutator obeying $[[A, B], A] = \mathbf{0} = [[A, B], B]$, one can express the exponential of the sum by a product of exponentials

$$e^{A+B} = e^A e^B e^{-[A,B]/2}. \tag{3.5}$$

If one now has many of these operators in the exponent $A_1, \ldots, A_n$ obeying $[A_i, A_j] = \alpha_{ij}\mathbf{1}$ such that $[[A_i, A_j], A_k] = \mathbf{0}$, one can generalize the above equation to

$$e^{\sum_{i=1}^{n} A_i} = e^{A_1} e^{A_2} \cdots e^{A_{n-1}} e^{A_n} e^{-\sum_{i<j}[A_i,A_j]/2}. \tag{3.6}$$

Exercise 3.2 (BCH formula) Show the generalization from Eq. (3.5) to Eq. (3.6).

Following the ideas in Ref. [3], we discretize the integral in the exponent of the time evolution operator:

$$\boldsymbol{U}(t) = \tau e^{-i \int_0^t \boldsymbol{H}_I(t')\,dt'} = \tau \lim_{\Delta t \to 0, N \to \infty} e^{\sum_{n=1}^{N} H_n \Delta t}, \tag{3.7}$$

where $H_n = -i\boldsymbol{H}_I(n\Delta t)$ with the constraint $N\Delta t = t$ remaining finite. Applying the generalized BCH formula (3.6), we obtain

$$\boldsymbol{U}(t) = \tau \prod_{n=1}^{N} e^{H_n \Delta t} e^{-\sum_{i<j}[H_i,H_j]/2} = \prod_{n=1}^{N} e^{H_n \Delta t} e^{-\sum_{i<j}[H_i,H_j]/2}, \tag{3.8}$$

where we note that the last exponential is just a number and that the operators are already time-ordered, such that the time ordering may simply be omitted. Recombining the exponentials of the operators, we see that the time ordering has no effect in this particular case:

$$U(t) = e^{-i\int_0^t H_I(t')\,dt'} = e^{\sigma^z \otimes \sum_k (\alpha_k(t)b_k - \alpha_k^*(t)b_k^\dagger)} \equiv e^{\sigma^z \otimes A(t)} \qquad (3.9)$$

with $\alpha_k(t) = (e^{-i\omega_k t} - 1)h_k/\omega_k$ and $A(t) = -A^\dagger(t)$.

Exercise 3.3 (Matrix exponentials) Show that for a unit vector $|n| = 1$ and a vector of Pauli matrices $\boldsymbol{\sigma} = (\sigma^x, \sigma^y, \sigma^z)$ one has

$$e^{(n\cdot\sigma)\otimes A} = \mathbf{1} \otimes \cosh(A) + (n \cdot \sigma) \otimes \sinh(A).$$

We can also write the unitary transformation as

$$U(t) = \mathbf{1} \otimes \frac{1}{2}\left(e^{+A(t)} + e^{-A(t)}\right) + \sigma^z \otimes \frac{1}{2}\left(e^{+A(t)} - e^{-A(t)}\right),$$
$$U^\dagger(t) = \mathbf{1} \otimes \frac{1}{2}\left(e^{+A(t)} + e^{-A(t)}\right) - \sigma^z \otimes \frac{1}{2}\left(e^{+A(t)} - e^{-A(t)}\right). \qquad (3.10)$$

When assuming an initial product state, the full density matrix is given by $\rho(t) = U(t)\rho_S^0 \otimes \bar{\rho}_B U^\dagger(t)$, which can be used to calculate any expectation value.

3.1.2 Reduced Dynamics

By performing the partial trace over the reservoir, we obtain the exact solution in the interaction picture:

$$\rho_S(t) = \text{Tr}_B\left\{U(t)\rho_S^0 \otimes \bar{\rho}_B U^\dagger(t)\right\}$$
$$= \rho_S^0 \frac{1}{4}\text{Tr}_B\left\{\left(e^{+2A(t)} + e^{-2A(t)} + 2\right)\bar{\rho}_B\right\}$$
$$- \rho_S^0 \sigma^z \frac{1}{4}\text{Tr}_B\left\{\left(e^{+2A(t)} - e^{-2A(t)}\right)\bar{\rho}_B\right\}$$
$$+ \sigma^z \rho_S^0 \frac{1}{4}\text{Tr}_B\left\{\left(e^{+2A(t)} - e^{-2A(t)}\right)\bar{\rho}_B\right\}$$
$$- \sigma^z \rho_S^0 \sigma^z \frac{1}{4}\text{Tr}_B\left\{\left(e^{+2A(t)} + e^{-2A(t)} - 2\right)\bar{\rho}_B\right\}, \qquad (3.11)$$

which can therefore be related to the expectation values $\langle e^{\pm 2A(t)}\rangle$ with respect to a thermal state. Since the bosonic annihilation and creation operators commute for different modes, we can separate the modes in the exponentials and write $\text{Tr}_B\{e^{2A(t)}\bar{\rho}_B\} = \prod_k T_k(t)$ with

$$T_k(t) = \text{Tr}_k\left\{e^{2\alpha_k(t)b_k - 2\alpha_k^* b_k^\dagger}\frac{e^{-\beta\omega_k b_k^\dagger b_k}}{Z_k}\right\}$$

$$= \sum_{n=0}^{\infty}\langle n|e^{-2\alpha_k^* b_k^\dagger}e^{+2\alpha_k(t)b_k}|n\rangle e^{-2|\alpha_k(t)|^2}e^{-\beta\omega_k n}\left[1 - e^{-\beta\omega_k}\right], \quad (3.12)$$

where we have used the BCH formula (3.5) and also inserted the normalized thermal state for mode k. For the matrix element we can use the identity

$$\langle n|e^{-\sigma^* b^\dagger}e^{\sigma b}|n\rangle = \mathscr{L}_n\big(|\sigma|^2\big), \tag{3.13}$$

with the Laguerre polynomial [4]

$$\mathscr{L}_n(x) = \frac{e^x}{n!}\frac{d^n}{dx^n}\big(e^{-x}x^n\big), \tag{3.14}$$

which further yields

$$T_k(t) = \sum_{n=0}^{\infty}\mathscr{L}_n\big(4|\alpha_k(t)|^2\big)e^{-2|\alpha_k(t)|^2}e^{-\beta\omega_k n}\left[1 - e^{-\beta\omega_k}\right]$$

$$= e^{-2|\alpha_k(t)|^2\coth(\beta\omega_k/2)}. \tag{3.15}$$

Therefore, we obtain for the sought-after expectation value

$$\text{Tr}_B\big\{e^{2A(t)}\bar{\rho}_B\big\} = e^{-2\sum_k |\alpha_k(t)|^2\coth(\beta\omega_k/2)} = \text{Tr}_B\big\{e^{-2A(t)}\bar{\rho}_B\big\}, \tag{3.16}$$

where the second equality sign follows from $A(t) = -A^\dagger(t)$ and the fact that the above expectation value is real. The exact solution for the system density matrix becomes

$$\rho_S(t) = \rho_S^0\frac{1}{2}\big[1 + e^{-2\sum_k |\alpha_k(t)|^2\coth(\beta\omega_k/2)}\big]$$

$$+ \sigma^z\rho_S^0\sigma^z\frac{1}{2}\big[1 - e^{-2\sum_k |\alpha_k(t)|^2\coth(\beta\omega_k/2)}\big], \tag{3.17}$$

which means that, as expected, the populations ρ_{00} and ρ_{11} are unaffected by the interaction with the reservoir, whereas the coherences evolve according to

$$\rho_{01}(t) = \rho_{01}^0 e^{-2\sum_k |\alpha_k(t)|^2\coth(\beta\omega_k/2)},$$

$$\rho_{10}(t) = \rho_{10}^0 e^{-2\sum_k |\alpha_k(t)|^2\coth(\beta\omega_k/2)}. \tag{3.18}$$

Inserting $|\alpha_k(t)|^2 = 2\frac{|h_k|^2}{\omega_k^2}[1 - \cos(\omega_k t)] = 4\frac{|h_k|^2}{\omega_k^2}\sin^2(\omega_k t/2)$, we eventually arrive at the well-known result that, in the pure dephasing model, the coherences decay as

$$\rho_{01}(t) = \exp\left\{-8\sum_k |h_k|^2 \frac{\sin^2(\omega_k t/2)}{\omega_k^2}\coth\left(\frac{\beta\omega_k}{2}\right)\right\}\rho_{01}^0, \tag{3.19}$$

which for a discrete spectrum of modes will display recurrences. Transforming to the continuum limit by introducing the spectral coupling density

$$J(\omega) = \sum_k |h_k|^2 \delta(\omega - \omega_k), \tag{3.20}$$

we note that as soon as $J(\omega)$ is represented as a smooth function, a popular choice being the parametrization [5]

$$J(\omega) = J_0 \frac{\omega^s}{\omega_{\text{ph}}^{1-s}} e^{-\omega/\omega_c}, \quad \text{for } \omega > 0, \tag{3.21}$$

the coherences will approach a vanishing stationary state $\lim_{t\to\infty}\rho_{01}(t) = 0$.

By performing a simple time derivative of the solution, one can now derive an exact master equation. For completeness we note here that this exact master equation has time-dependent rates. In addition, it is not of Lindblad form (also for constant time) but must—since the solution is exact—nevertheless preserve positivity of the density matrix.

In general, the speed of decoherence depends on the temperature and coupling strength, etc. For high temperatures, we can expand the integrand and solve the special case $s = 1$ and $\omega_c \to \infty$ in the above parametrization explicitly:

$$\rho_{01}(t) \approx e^{-4\pi\frac{J_0}{\beta}t}\rho_{01}^0. \tag{3.22}$$

This result can also be reproduced within a master equation approach, as described below.

3.1.3 Master Equation Approach

Identifying a single system and bath coupling operator in the interaction Hamiltonian $A = \sigma^z$ and $B = \sum_k(h_k b_k + h_k^* b_k^\dagger)$, respectively, we first calculate the bath correlation function

$$C(t) = \langle B(t)B\rangle = \sum_{kk'}\langle\left(h_k b_k e^{-i\omega_k t} + h_k^* b_k^\dagger e^{+i\omega_k t}\right)\left(h_{k'} b_{k'} + h_{k'}^* b_{k'}^\dagger\right)\rangle$$

$$= \sum_k |h_k|^2\left\{e^{-i\omega_k t}[1 + n_B(\omega_k)] + e^{+i\omega_k t}n_B(\omega_k)\right\}$$

$$= \int_0^\infty J(\omega)\left\{e^{-i\omega t}\left[1 + n_B(\omega)\right] + e^{+i\omega t} n_B(\omega)\right\} d\omega$$

$$= \int_{-\infty}^{+\infty} J(\omega) e^{-i\omega t} J(\omega)\left[1 + n_B(\omega)\right] d\omega, \tag{3.23}$$

where we have analytically continued the spectral coupling density to negative frequencies $J(-\omega) = -J(\omega)$. This enables us to identify the Fourier transform of the correlation function as

$$\gamma(\omega) = 2\pi J(\omega)\left[1 + n_B(\omega)\right]. \tag{3.24}$$

With the help of Eq. (2.38) this can be used to calculate the odd Fourier transform numerically. The quantum optical master equation in Definition 2.3 then yields

$$\dot{\rho}_{00} = \dot{\rho}_{11} = 0,$$

$$\dot{\rho}_{01} = -i(E_0 - E_1 + \sigma_{00} - \sigma_{11})\rho_{01} + \left(\gamma_{00,11} - \frac{1}{2}\gamma_{00,00} - \frac{1}{2}\gamma_{11,11}\right)\rho_{01} \tag{3.25}$$

$$= -i(E_0 - E_1 + \sigma_{00} - \sigma_{11})\rho_{01} - 2\gamma(0)\rho_{01}.$$

The first two equations just express the fact that the interaction does not change the system energy, which is also obeyed by the master equation solution.

The Lamb-shift terms can be expressed with the odd Fourier transform of the reservoir correlation function $\sigma_{00} = \sigma(0)/(2i) = \sigma_{11}$, and thus they cancel in the evolution of the coherences. Therefore, we obtain for the coherences a decay according to $\rho_{01}(t) = e^{-i(E_0-E_1)t} e^{-2\gamma(0)t}|\rho_{01}^0|$. The first exponential can be transformed away by switching to the interaction picture $\rho_S(t) = e^{+iH_S t}\rho_S(t)e^{-iH_S t}$, where one only has $\rho_{01}(t) = e^{-2\gamma(0)t}|\rho_{01}^0|$. Now, assuming high temperatures and an ohmic spectral coupling density $J(\omega) = J_0\omega$, the limit becomes $\lim_{\omega\to 0}\gamma(0) = 2\pi J_0/\beta$, which perfectly coincides with the result in Eq. (3.22).

We finally note that the Lindblad form only guarantees positivity of the solution if initialized with a valid, i.e., positive, density matrix.

3.2 Quantum Dot Coupled to Two Fermionic Leads

As one of the simplest fermionic models, we consider a single electron transistor (SET). The system, bath, and interaction Hamiltonians are given by

$$H_S = \varepsilon d^\dagger d, \qquad H_B = \sum_k \varepsilon_{kL} c_{kL}^\dagger c_{kL} + \sum_k \varepsilon_{kR} c_{kR}^\dagger c_{kR},$$

$$H_I = \sum_k \left(t_{kL} d c_{kL}^\dagger + t_{kL}^* c_{kL} d^\dagger\right) + \sum_k \left(t_{kR} d c_{kR}^\dagger + t_{kR}^* c_{kR} d^\dagger\right), \tag{3.26}$$

where d is a fermionic annihilation operator on the dot and $c_{k\alpha}$ are fermionic annihilation operators of an electron in the kth mode of lead α. Obviously, this corresponds to a quadratic fermionic Hamiltonian, which can in principle be solved exactly by various methods, such as non-equilibrium Green's functions [6] or even the equation of motion approach [7]. Such quadratic models are useful for studying exact transport properties [8] or exact master equations [9].

3.2.1 Heisenberg Picture Dynamics

To be as self-contained as possible, here we simply compute the Heisenberg equations of motion for the system and bath annihilation operators (we denote operators in the Heisenberg picture by boldface symbols):

$$\dot{\boldsymbol{d}} = -i\varepsilon\boldsymbol{d} + i\sum_k \left[t_{kL}^* \boldsymbol{c}_{kL} + t_{kR}^* \boldsymbol{c}_{kR}\right],$$

$$\dot{\boldsymbol{c}}_{kL} = -i\varepsilon_{kL}\boldsymbol{c}_{kL} + it_{kL}\boldsymbol{d}, \qquad (3.27)$$

$$\dot{\boldsymbol{c}}_{kR} = -i\varepsilon_{kR}\boldsymbol{c}_{kR} + it_{kR}\boldsymbol{d}.$$

Surprisingly, this system is already closed, and we obtain its solution by performing a Laplace transform [10]:

$$z\tilde{d}(z) - d = -i\varepsilon\tilde{d}(z) + i\sum_k \left[t_{kL}^* \tilde{c}_{kL}(z) + t_{kR}^* \tilde{c}_{kR}(z)\right],$$

$$z\tilde{c}_{kL}(z) - c_{kL} = -i\varepsilon_{kL}\tilde{c}_{kL}(z) + it_{kL}\tilde{d}(z), \qquad (3.28)$$

$$z\tilde{c}_{kR}(z) - c_{kR} = -i\varepsilon_{kR}\tilde{c}_{kR}(z) + it_{kR}\tilde{d}(z).$$

In the above equations, we can eliminate the operators $\tilde{c}_{kL}(z)$ and $\tilde{c}_{kR}(z)$. This yields for the dot annihilation operator

$$\tilde{d}(z) = \frac{d + i\sum_k \left(\frac{t_{kL}^* c_{kL}}{z + i\varepsilon_{kL}} + \frac{t_{kR}^* c_{kR}}{z + i\varepsilon_{kR}}\right)}{z + i\varepsilon + \sum_k \left(\frac{|t_{kL}|^2}{z + i\varepsilon_{kL}} + \frac{|t_{kR}|^2}{z + i\varepsilon_{kR}}\right)}$$

$$\equiv \tilde{f}(z)d + \sum_k \left(\tilde{g}_{kL}(z)c_{kL} + \tilde{g}_{kR}(z)c_{kR}\right), \qquad (3.29)$$

where we have introduced the functions $\tilde{g}_{k\alpha}(z)$ and $\tilde{f}(z)$. This expression also yields the solution for the operators of the right lead modes,

$$\tilde{c}_{k\alpha}(z) = \frac{1}{z + i\varepsilon_{k\alpha}} c_{k\alpha} + \frac{it_{k\alpha}}{z + i\varepsilon_{k\alpha}} \tilde{d}(z). \qquad (3.30)$$

Inverting the Laplace transform may now be achieved by identifying the poles and applying the residue theorem. In the wide-band limit discussed below, this becomes particularly simple.

3.2.2 Stationary Occupation

The time-dependent occupation $n(t) = \langle d^\dagger(t) d(t) \rangle$ is found by inverting the Laplace transform. For the moment we do it formally and determine the expectation value

$$
\begin{aligned}
n(t) &= \left\langle \left[f^*(t) d^\dagger + \sum_k (g_{kL}^*(t) c_{kL}^\dagger + g_{kR}^*(t) c_{kR}^\dagger) \right] \right. \\
&\quad \left. \times \left[f(t) d + \sum_k (g_{kL}(t) c_{kL} + g_{kR}(t) c_{kR}) \right] \right\rangle \\
&= |f(t)|^2 n_0 + \sum_k (|g_{kL}(t)|^2 f_L(\varepsilon_{kL}) + |g_{kR}(t)|^2 f_R(\varepsilon_{kR})), \quad (3.31)
\end{aligned}
$$

where we have used a product state as an initial one,

$$
\rho_0 = \rho_S^0 \frac{e^{-\beta_L (H_L - \mu_L N_L)}}{Z_L} \frac{e^{-\beta_R (H_R - \mu_R N_R)}}{Z_R} \tag{3.32}
$$

with the lead Hamiltonians $H_\alpha = \sum_k \varepsilon_{k\alpha} c_{k\alpha}^\dagger c_{k\alpha}$ and the lead particle numbers $N_\alpha = \sum_k c_{k\alpha}^\dagger c_{k\alpha}$. These eventually yield the only nonvanishing expectation values $n_0 = \langle d^\dagger d \rangle$ and $f_\alpha(\varepsilon_{k\alpha}) = \langle c_{k\alpha}^\dagger c_{k\alpha} \rangle$. Inverse lead temperatures β_α and chemical potentials μ_α thereby only enter implicitly in the Fermi functions. Therefore, to find the exact solution for the time-dependent dot occupation, we have to find the inverse Laplace transform of the following:

$$
\begin{aligned}
\tilde{f}(z) &= \frac{1}{z + i\varepsilon + \sum_k \left(\frac{|t_{kL}|^2}{z + i\varepsilon_{kL}} + \frac{|t_{kR}|^2}{z + i\varepsilon_{kR}} \right)}, \\
\tilde{g}_{k\alpha}(z) &= \frac{i t_{k\alpha}^*}{[z + i\varepsilon_{k\alpha}][z + i\varepsilon + \sum_k \left(\frac{|t_{kL}|^2}{z + i\varepsilon_{kL}} + \frac{|t_{kR}|^2}{z + i\varepsilon_{kR}} \right)]},
\end{aligned}
\tag{3.33}
$$

which heavily depends on the number of modes and their distribution in the reservoir. For example, any system with a finite number of reservoir modes will exhibit recurrences to the initial state.

Only systems with a continuous spectrum of reservoir modes can be expected to yield a stationary system state. To obtain that limit, for simplicity we assume $N + 1$ modes in each reservoir, $-N/2 \leq k \leq +N/2$. These are distributed over the energies as $\varepsilon_{k\alpha} = k\Omega/\sqrt{N}$ and are assumed to couple more weakly to the dot as their momentum increases:

$$|t_{k\alpha}|^2 = \frac{\Omega}{2\pi\sqrt{N}} \frac{\Gamma_\alpha \delta_\alpha^2}{(k\Omega/\sqrt{N})^2 + \delta_\alpha^2}. \tag{3.34}$$

Letting the number of reservoir modes N go to infinity, we can replace the summation in the denominators by a continuous integral:

$$\tilde{f}(z) \approx \frac{1}{z + i\varepsilon + \int \frac{1}{2\pi}(\frac{\Gamma_L \delta_L^2}{\omega^2 + \delta_L^2} + \frac{\Gamma_R \delta_R^2}{\omega^2 + \delta_R^2}) \frac{1}{z+i\omega} d\omega} = \frac{1}{z + i\varepsilon + \frac{1}{2}(\frac{\Gamma_L \delta_L}{z+\delta_L} + \frac{\Gamma_R \delta_R}{z+\delta_R})},$$

$$\tilde{g}_{k\alpha}(z) \approx \frac{it_{k\alpha}^*}{(z + i\varepsilon_{k\alpha})[z + i\varepsilon + \int \frac{1}{2\pi}(\frac{\Gamma_L \delta_L^2}{\omega^2 + \delta_L^2} + \frac{\Gamma_R \delta_R^2}{\omega^2 + \delta_R^2}) \frac{1}{z+i\omega} d\omega]}$$

$$= \frac{1}{[z + i\varepsilon_{k\alpha}][z + i\varepsilon + \frac{1}{2}(\frac{\Gamma_L \delta_L}{z+\delta_L} + \frac{\Gamma_R \delta_R}{z+\delta_R})]}. \tag{3.35}$$

We note that this transfer from a discrete to a continuous spectrum of reservoir modes is commonly performed formally by introducing the energy-dependent tunneling rates

$$\Gamma_\alpha(\omega) = 2\pi \sum_k |t_{k\alpha}|^2 \delta(\omega - \varepsilon_{k\alpha}). \tag{3.36}$$

Here, we have thereby assumed a Lorentzian-shaped tunneling rate [11]

$$\Gamma_\alpha(\omega) = \frac{\Gamma_\alpha \delta_\alpha^2}{\omega^2 + \delta_\alpha^2}. \tag{3.37}$$

The simple pole structure of these tunneling rates renders analytic calculations simple. Superpositions of many Lorentzian shapes with shifted centers may approximate quite general tunneling rates [12].

To obtain sufficiently simple results, we assume the wide-band limit $\delta_\alpha \to \infty$ (within which the tunneling rates are flat), where one obtains the simple expression

$$\tilde{f}(z) \to \frac{1}{z + i\varepsilon + (\Gamma_L + \Gamma_R)/2},$$

$$\tilde{g}_{k\alpha}(z) \to \frac{it_{k\alpha}^*}{(z + i\varepsilon_{k\alpha})[z + i\varepsilon + (\Gamma_L + \Gamma_R)/2]}. \tag{3.38}$$

Inserting the inverse Laplace transforms of these expressions,

$$f(t) \rightarrow e^{-i\varepsilon t} e^{-\Gamma t/2},$$

$$g_{k\alpha}(t) \rightarrow \frac{t_{k\alpha}^*(e^{-i\varepsilon t}e^{-\Gamma t/2} - e^{-i\varepsilon_{k\alpha}t})}{\varepsilon_{k\alpha} - \varepsilon + i\Gamma/2} \tag{3.39}$$

(with $\Gamma \equiv \Gamma_L + \Gamma_R$) into Eq. (3.31), we obtain by switching to a continuum representation

$$n(t) = e^{-\Gamma t}n_0 + \sum_k \sum_\alpha |t_{k\alpha}|^2 f_\alpha(\varepsilon_{k\alpha})4\frac{1 - 2e^{-\Gamma t/2}\cos[(\varepsilon_{k\alpha} - \varepsilon)t] + e^{-\Gamma t}}{\Gamma^2 + 4(\varepsilon_{k\alpha} - \varepsilon)^2}$$

$$= e^{-\Gamma t}n_0 + \sum_\alpha \int d\omega \Gamma_\alpha f_\alpha(\omega)\frac{4}{2\pi}\frac{1 - 2e^{-\Gamma t/2}\cos[(\omega - \varepsilon)t] + e^{-\Gamma t}}{\Gamma^2 + 4(\omega - \varepsilon)^2}.$$

$$\tag{3.40}$$

The long-term limit can, because $\Gamma \geq 0$, be read off easily, and the stationary occupation becomes

$$\bar{n} = \sum_\alpha \int d\omega \, \Gamma_\alpha f_\alpha(\omega)\frac{2}{\pi}\frac{1}{\Gamma^2 + 4(\omega - \varepsilon)^2}. \tag{3.41}$$

With the above formula for the stationary occupation valid for the wide-band limit, one can easily demonstrate the following.

At infinite bias $f_L(\omega) = 1$ and $f_R(\omega) = 0$, the stationary occupation approaches $\bar{n} \rightarrow \Gamma_L/(\Gamma_L + \Gamma_R)$, regardless of the coupling strength. A similar result is of course obtained for reverse infinite bias where $\bar{n} \rightarrow \Gamma_R/(\Gamma_L + \Gamma_R)$.

When the quantum dot is coupled weakly to a single bath only (e.g., $\Gamma_R(\omega) = 0$), the stationary occupation approaches the Fermi distribution of the coupled lead, evaluated at the dot energy (e.g., $\bar{n} = f_L(\varepsilon) + \mathcal{O}\{\Gamma_L\}$). This implies that, for weak coupling to an equilibrium reservoir, the system will equilibrate with the temperature and chemical potential of the reservoir, consistent with what one expects from a master equation approach.

When the dot is coupled weakly to both reservoirs, the stationary state approaches

$$\bar{n} \rightarrow \frac{\Gamma_L f_L(\varepsilon) + \Gamma_R f_R(\varepsilon)}{\Gamma_L + \Gamma_R}, \tag{3.42}$$

which is also obtained within a master equation approach (compare Sect. 5.1).

Exercise 3.4 (Weak coupling limit) Show that Eq. (3.41) reduces in the weak-coupling limit to Eq. (3.42) by using a representation of the Dirac delta distribution,

$$\delta(x) = \lim_{\varepsilon \to 0} \frac{1}{\pi}\frac{\varepsilon}{x^2 + \varepsilon^2}.$$

In contrast, for the strong coupling limit, the stationary occupation will be suppressed, $\bar{n} \to 0$, as the exact solution for the stationary state is no longer localized on the dot.

3.2.3 Stationary Current

The stationary current from left to right through the SET can be defined as the long-term limit of the change of particle numbers at the right lead,

$$I = \lim_{t \to \infty} \frac{d}{dt} \left\langle \sum_k c_{kR}^\dagger c_{kR} \right\rangle, \tag{3.43}$$

which we can evaluate in the Heisenberg picture as we did for the stationary occupation. Using Eq. (3.30), the right lead modes can be written as

$$\tilde{c}_{kR}(z) = \frac{it_{kR}}{(z + i\varepsilon_{kR})(z + i\varepsilon + \Gamma/2)} d + \frac{1}{z + i\varepsilon_{kR}} c_{kR}$$

$$- \sum_q \frac{t_{kR} t_{qL}^*}{(z + i\varepsilon_{kR})(z + i\varepsilon_{qL})(z + i\varepsilon + \Gamma/2)} c_{qL}$$

$$- \sum_q \frac{t_{kR} t_{qR}^*}{(z + i\varepsilon_{kR})(z + i\varepsilon_{qR})(z + i\varepsilon + \Gamma/2)} c_{qR}. \tag{3.44}$$

Now, performing the inverse Laplace transform and neglecting all transient dynamics, we obtain the asymptotic evolution of the annihilation operators in the Heisenberg picture:

$$c_{kR}(t) \to \left(-\frac{t_{kR} e^{-i\varepsilon_{kR}t}}{\varepsilon_{kR} - \varepsilon + i\Gamma/2} \right) d + e^{-i\varepsilon_{kR}t} c_{kR}$$

$$+ \sum_q \frac{t_{kR} t_{qL}^*}{\varepsilon_{kR} - \varepsilon_{qL}} \left(\frac{e^{-i\varepsilon_{qL}t}}{\varepsilon_{qL} - \varepsilon + i\Gamma/2} - \frac{e^{-i\varepsilon_{kR}t}}{\varepsilon_{kR} - \varepsilon + i\Gamma/2} \right) c_{qL}$$

$$+ \sum_q \frac{t_{kR} t_{qR}^*}{\varepsilon_{kR} - \varepsilon_{qR}} \left(\frac{e^{-i\varepsilon_{qR}t}}{\varepsilon_{qR} - \varepsilon + i\Gamma/2} - \frac{e^{-i\varepsilon_{kR}t}}{\varepsilon_{kR} - \varepsilon + i\Gamma/2} \right) c_{qR}. \tag{3.45}$$

The occupation of the right lead therefore becomes

$$N_R \to \sum_k \frac{|t_{kR}|^2}{(\varepsilon_{kR} - \varepsilon)^2 + \Gamma^2/4} n_0 + N_R^0$$

$$- \sum_{kq} \left[\frac{t_{kR} t_{qR}^*}{\varepsilon_{kR} - \varepsilon_{qR}} e^{+i\varepsilon_{kR}t} \right.$$

$$\times \left(\frac{e^{-i\varepsilon_{qR}t}}{\varepsilon_{qR} - \varepsilon + i\Gamma/2} - \frac{e^{-i\varepsilon_{kR}t}}{\varepsilon_{kR} - \varepsilon + i\Gamma/2} \right) \delta_{kq} f_R(\varepsilon_{kR}) + \text{h.c.} \Bigg]$$

$$+ \sum_{kq} \frac{|t_{kR}|^2 |t_{qL}|^2}{(\varepsilon_{kR} - \varepsilon_{qL})^2} \left(\frac{e^{+i\varepsilon_{qL}t}}{\varepsilon_{qL} - \varepsilon - i\Gamma/2} - \frac{e^{+i\varepsilon_{kR}t}}{\varepsilon_{kR} - \varepsilon - i\Gamma/2} \right)$$

$$\times \left(\frac{e^{-i\varepsilon_{qL}t}}{\varepsilon_{qL} - \varepsilon + i\Gamma/2} - \frac{e^{-i\varepsilon_{kR}t}}{\varepsilon_{kR} - \varepsilon + i\Gamma/2} \right) f_L(\varepsilon_{qL})$$

$$+ \sum_{kq} \frac{|t_{kR}|^2 |t_{qR}|^2}{(\varepsilon_{kR} - \varepsilon_{qR})^2} \left(\frac{e^{+i\varepsilon_{qR}t}}{\varepsilon_{qR} - \varepsilon - i\Gamma/2} - \frac{e^{+i\varepsilon_{kR}t}}{\varepsilon_{kR} - \varepsilon - i\Gamma/2} \right)$$

$$\times \left(\frac{e^{-i\varepsilon_{qR}t}}{\varepsilon_{qR} - \varepsilon + i\Gamma/2} - \frac{e^{-i\varepsilon_{kR}t}}{\varepsilon_{kR} - \varepsilon + i\Gamma/2} \right) f_R(\varepsilon_{qR}). \tag{3.46}$$

The first term is just triggered by the initial occupation of the dot, and the second term corresponds to the initial occupation of the right lead. These terms are constant and cannot contribute to the current, which however is different for all other terms. Introducing the tunneling rates in the wide-band limit $\Gamma_\alpha \approx \Gamma_\alpha(\omega) = \sum_k |t_{k\alpha}|^2 \delta(\omega - \varepsilon_{k\alpha})$, we can represent the right lead occupation by integrals:

$$N_R \rightarrow \frac{1}{2\pi} \int d\omega \, \frac{\Gamma_R}{(\omega - \varepsilon)^2 + \Gamma^2/4} n_0$$

$$+ N_R^0 - \frac{1}{2\pi} \int d\omega \, \Gamma_R f_R(\omega) \left[\frac{4 + 4i\omega t - 2t(\Gamma + 2i\varepsilon)}{(2\omega + i\Gamma - 2\varepsilon)^2} + \text{h.c.} \right]$$

$$+ \frac{1}{4\pi^2} \int d\omega \, d\omega' \left(\Gamma_L \Gamma_R f_L(\omega') + \Gamma_R^2 f_R(\omega') \right) \frac{1}{(\omega - \omega')^2}$$

$$\times \left| \frac{e^{-i\omega' t}}{\omega' - \varepsilon + i\Gamma/2} - \frac{e^{-i\omega t}}{\omega - \varepsilon + i\Gamma/2} \right|^2. \tag{3.47}$$

Whereas the first two terms are constant and do not contribute to the current, all other terms yield a nonvanishing contribution. The long-term limit of the time derivative of the very last term is a bit involved to determine. It can be found, e.g., by using properties of the Laplace transform. To evaluate the current, we therefore consider the limit

$$F(\omega') \equiv \lim_{t \to \infty} \frac{d}{dt} \int d\omega \, \frac{1}{(\omega - \omega')^2} \left| \frac{e^{-i\omega' t}}{\omega' - \varepsilon + i\Gamma/2} - \frac{e^{-i\omega t}}{\omega - \varepsilon + i\Gamma/2} \right|^2$$

$$= \lim_{z \to 0} z \int_0^\infty dt \, e^{-zt} \frac{d}{dt} \int d\omega \frac{1}{(\omega - \omega')^2} \left| \frac{e^{-i\omega' t}}{\omega' - \varepsilon + i\Gamma/2} - \frac{e^{-i\omega t}}{\omega - \varepsilon + i\Gamma/2} \right|^2$$

$$= \frac{8\pi}{\Gamma^2 + 4(\omega' - \varepsilon)^2}, \tag{3.48}$$

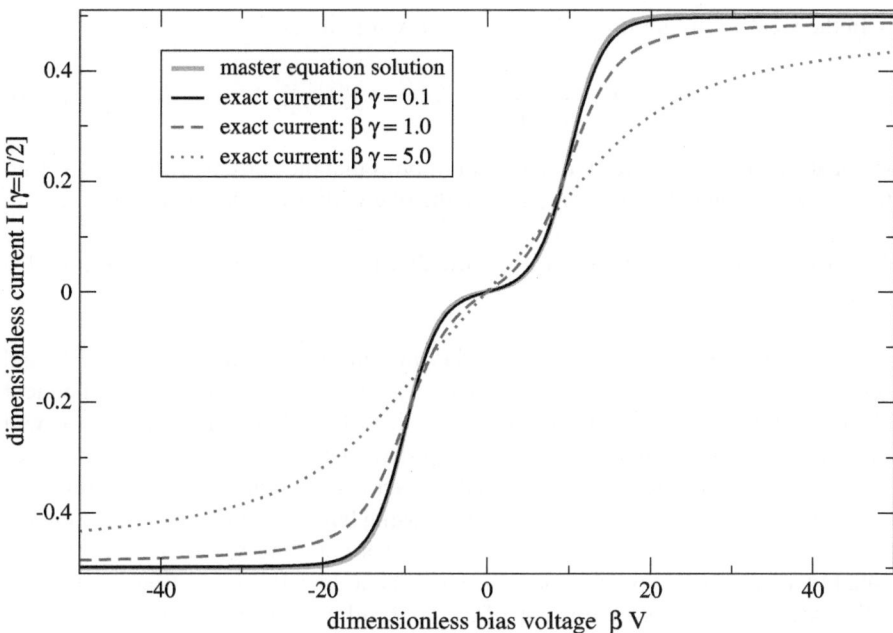

Fig. 3.1 Plot of the electronic matter current (in units of $\gamma = \Gamma_L = \Gamma_R = \Gamma/2$) versus the bias voltage for symmetric tunneling rates and equal electronic temperatures $\beta_L = \beta_R = \beta$ and dot level $\beta\varepsilon = 5$. For a small coupling strength, the exact (*black solid*) and master equation (*brown bold*) solutions coincide for all bias voltages. For stronger couplings (*red dashed* and *green dotted*, respectively), the determination of the dot level ε from steps in the current is no longer possible (Color figure online)

which with its Lorentzian shape converges for small Γ towards a Dirac delta distribution. The current becomes

$$
\begin{aligned}
I &= -\frac{1}{\pi}\int d\omega\,\Gamma_R f_R(\omega)\frac{\Gamma/2}{(\omega-\varepsilon)^2+(\Gamma/2)^2}\\
&\quad + \frac{1}{\pi\Gamma}\int d\omega\big(\Gamma_L\Gamma_R f_L(\omega)+\Gamma_R^2 f_R(\omega)\big)\frac{\Gamma/2}{(\omega-\varepsilon)^2+(\Gamma/2)^2}\\
&= \frac{\Gamma_L\Gamma_R}{\Gamma_L+\Gamma_R}\int d\omega\big[f_L(\omega)-f_R(\omega)\big]\frac{1}{\pi}\frac{\Gamma/2}{(\omega-\varepsilon)^2+(\Gamma/2)^2}.
\end{aligned}
\tag{3.49}
$$

Alternatively, this expression can also be derived by evaluating the expectation value of the current operator directly $I = i\sum_k t_{kR}\langle c_{kR}^\dagger(t)d(t)\rangle + \text{h.c.}$ The integrals in the above expression can be solved analytically by analysis in the complex plane, but here we will be content with the above integral representation, which can also be found using non-equilibrium Green's functions [6]. For consistency, we note that the current is antisymmetric under the exchange of left and right leads as expected.

In the weak coupling limit $\Gamma \to 0$, the current reduces to

$$I = \frac{\Gamma_L \Gamma_R}{\Gamma_L + \Gamma_R} \left[f_L(\varepsilon) - f_R(\varepsilon) \right], \tag{3.50}$$

which at equal temperatures left and right implies that the current always flows from the lead with larger chemical potential to the one with lower chemical potential.

Exercise 3.5 (Weak coupling limit) Show that Eq. (3.50) follows from Eq. (3.49) when $\Gamma \to 0$.

Finally, we note further that, in the infinite bias limit ($f_L(\omega) \to 1$ and $f_R(\omega) \to 0$), the current becomes $I = \Gamma_L \Gamma_R / (\Gamma_L + \Gamma_R)$, which is independent of the coupling strength and also consistent with Eq. (3.50). In Sect. 5.1 we will find that the master equation approach applied to the same problem reproduces Eq. (3.50) and therefore coincides with the exact result in the infinite bias limit.

Figure 3.1 demonstrates the effect of increasing but symmetric coupling strengths $\Gamma_L = \Gamma_R = \gamma$ on the current. Whereas the weak coupling result is well approximated when $\beta \gamma \ll 1$, one may observe significant deviations for strong couplings. In the example shown, spectroscopy of the dot level ε via detecting steps in the I–V characteristics is therefore only possible in the weak coupling limit.

References

1. W.G. Unruh, Maintaining coherence in quantum computers. Phys. Rev. A **51**, 992 (1995)
2. J.H. Reina, L. Quiroga, N.F. Johnson, Decoherence of quantum registers. Phys. Rev. A **65**, 032326 (2002)
3. D.A. Lidar, Z. Bihary, K.B. Whaley, From completely positive maps to the quantum Markovian semigroup master equation. Chem. Phys. **268**, 35 (2001)
4. M. Abramowitz, I.A. Stegun (eds.), Handbook of Mathematical Functions (National Bureau of Standards, 1970)
5. T. Brandes, Coherent and collective quantum optical effects in mesoscopic systems. Phys. Rep. **408**, 315 (2005)
6. H. Haug, A.-P. Jauho, *Quantum Kinetics in Transport and Optics of Semiconductors* (Springer, Berlin, 2008)
7. G. Schaller, P. Zedler, T. Brandes, Systematic perturbation theory for dynamical coarse-graining. Phys. Rev. A **79**, 032110 (2009)
8. A. Dhar, K. Saito, P. Hänggi, Nonequilibrium density-matrix description of steady-state quantum transport. Phys. Rev. E **85**, 011126 (2012)
9. L.-P. Yang, C.Y. Cai, D.Z. Xu, W.-M. Zhang, C.P. Sun, Master equation and dispersive probing of a non-Markovian process. Phys. Rev. A **87**, 012110 (2013)
10. G.B. Arfken, H.J. Weber, *Mathematical Methods for Physicists* (Elsevier, Oxford, 2005)
11. P. Zedler, G. Schaller, G. Kießlich, C. Emary, T. Brandes, Weak coupling approximations in non-Markovian transport. Phys. Rev. B **80**, 045309 (2009)
12. U. Kleinekathöfer, Non-Markovian theories based on a decomposition of the spectral density. J. Chem. Phys. **121**, 2505 (2004)

Chapter 4
Technical Tools

Abstract This chapter provides tools that are useful for the solution and handling of master equations. We start with simple analytic approaches including the equation of motion technique and the quantum regression theorem. As numerical techniques, we investigate a Runge–Kutta solver applied to a master equation and introduce the stochastic Schrödinger equation. For rate equations obeying local detailed balance, we treat the evolution of the Shannon entropy and connect it to the full counting statistics. We show how the statistics of energy and matter transfers can be extracted from the master equation. In particular, we demonstrate how the moments and cumulants of the corresponding distributions can be obtained. Finally, we relate symmetries in the respective generating functions with the fluctuation theorem for entropy production. The methods in this chapter may also be applied to Markovian master equations that are not in Lindblad form; only constant coefficients and a time-local evolution equation for the density matrix are required.

4.1 Analytic Techniques for Solving Master Equations

Trivially, as the superoperator notation in Sect. 1.6 allows us to write master equations as systems of ordinary coupled differential equations with constant coefficients, we may obtain the solution of the master equation by exponentiating the Liouvillian superoperator

$$\rho(t) = e^{\mathscr{L}t}\rho_0. \tag{4.1}$$

This is usually quite difficult and constrained to very small dimensions of $\mathscr{L}$. In addition, since the Liouville superoperator $\mathscr{L}$ is not hermitian, it need not even have a spectral decomposition.

Exercise 4.1 (Single resonant level) Calculate the matrix exponential of the Liouvillian superoperator for a single resonant level tunnel-coupled to a single junction

$$\mathscr{L} = \begin{pmatrix} -\Gamma f & +\Gamma(1-f) \\ +\Gamma f & -\Gamma(1-f) \end{pmatrix}$$

G. Schaller, *Open Quantum Systems Far from Equilibrium*, Lecture Notes in Physics 881, 61
DOI 10.1007/978-3-319-03877-3_4,
© Springer International Publishing Switzerland 2014

when the dot level is much lower than the Fermi edge ($f \to 1$) and when it is much larger than the Fermi edge $f \to 0$.

Thus, solving the master equation by brute force is usually not advisable.

4.1.1 Laplace Transform

If one is only interested in stationary properties, it is often useful to obtain the formal solution by performing a Laplace transform, $\tilde{\rho}(z) = \int_0^\infty \rho(t)e^{-zt}\,dt$. In frequency space, the master equation is then reduced to an algebraic problem, which may readily be solved by

$$\tilde{\rho}(z) = \frac{1}{z \cdot \mathbf{1} - \mathscr{L}}\rho_0, \tag{4.2}$$

where ρ_0 is just the initial state. This just requires the computation of the inverse of $z \cdot \mathbf{1} - \mathscr{L}$, which is significantly less demanding than exponentiating a matrix. The main obstacle however is the calculation of the inverse Laplace transform, which requires one to identify the poles of $[z \cdot \mathbf{1} - \mathscr{L}]^{-1}$. In cases where one is only interested in stationary values, it can be useful to compute the steady-state values of observables by exploiting properties of the Laplace transform: the long-time limit of a function in the time domain can be obtained from a small-z limit in the frequency domain, $\lim_{t\to\infty} f(t) = \lim_{z\to 0} z\tilde{f}(z)$. Applied to an observable, this yields

$$\langle \bar{A} \rangle = \mathrm{Tr}\{A\bar{\rho}\} = \lim_{z\to 0} z\,\mathrm{Tr}\{A\tilde{\rho}(z)\} = \lim_{z\to 0} z\,\mathrm{Tr}\left\{A\frac{1}{z \cdot \mathbf{1} - \mathscr{L}}\rho_0\right\}, \tag{4.3}$$

such that the trace can be performed in frequency space, which may sometimes yield significant simplifications.

4.1.2 Equation of Motion Technique

Instead of solving the master equation for the density matrix, it may be more favorable to derive a related linear set of first-order differential equations for observables $\langle B_i \rangle(t)$ of interest instead. In fact, for infinitely large system Hilbert space dimensions such a procedure might even be necessary:

$$\langle \dot{B}_i(t) \rangle = \mathrm{Tr}\{B_i \mathscr{L}\rho(t)\}$$

$$= -i\,\mathrm{Tr}\{B_i[H, \rho(t)]\} + \sum_\alpha \gamma_\alpha \,\mathrm{Tr}\left\{B_i\left(L_\alpha \rho(t) L_\alpha^\dagger - \frac{1}{2}\{L_\alpha^\dagger L_\alpha, \rho(t)\}\right)\right\}$$

$$= \mathrm{Tr}\left\{\left(+\mathrm{i}[H, B_i] + \sum_\alpha \gamma_\alpha \left[L_\alpha^\dagger B_i L_\alpha - \frac{1}{2}\{L_\alpha^\dagger L_\alpha, B_i\}\right]\right)\rho(t)\right\}$$

$$= \mathrm{Tr}\left\{\left[\sum_j G_{ij} B_j\right]\rho(t)\right\} = \sum_j G_{ij}\langle B_j(t)\rangle, \tag{4.4}$$

where in the last line we have used the fact that there is for a finite-dimensional system Hilbert space only a finite set of linearly independent operators. The linear coefficients G_{ij} have to be found for each master equation separately. The advantage is that, for well-chosen sets of operators, one can hope to end up with a much smaller set of equations than are necessary for solving the complete master equation. For example, this is the case when the matrix G_{ij} has a block structure.

Exercise 4.2 (Equation of motion for the harmonic oscillator) Calculate the expectation value of $a + a^\dagger$ for a cavity in a vacuum bath

$$\dot{\rho} = -\mathrm{i}[H, \rho] + \gamma\left[a\rho a^\dagger - \frac{1}{2}\{a^\dagger a, \rho\}\right]. \tag{4.5}$$

4.1.3 Quantum Regression Theorem

As with the Heisenberg picture for closed quantum systems, it may be favorable to keep the density matrix as constant and to shift the complete time dependence to the operators. From Eq. (4.4) we can conclude for the operators that

$$\dot{B}_i(t) = \mathscr{L}^\dagger B_i(t) = +\mathrm{i}[H, B_i(t)] + \sum_\alpha \gamma_\alpha \left[L_\alpha^\dagger B_i(t) L_\alpha - \frac{1}{2}\{L_\alpha^\dagger L_\alpha, B_i(t)\}\right]$$

$$= \sum_j G_{ij} B_j(t), \tag{4.6}$$

where we have introduced the adjoint Liouvillian $\mathscr{L}^\dagger$. For open quantum systems, it is however often important to calculate the expectation values of operators at different times, which may be facilitated with the help of the quantum regression theorem. We find directly from properties of the matrix exponential that

$$\frac{d}{d\tau} B_i(t + \tau) = \mathscr{L}^\dagger B_i(t + \tau) = \sum_j G_{ij} B_j(t + \tau). \tag{4.7}$$

Using this relation, we find the quantum regression theorem for two-point correlation functions.

Definition 4.1 (Quantum regression) Let single observables follow the closed equation $\langle \dot{B}_i \rangle = \sum_j G_{ij}\langle B_j \rangle$. Then, the two-point correlation functions obey the equations

$$\frac{d}{d\tau}\langle B_i(t+\tau)B_\ell(t)\rangle = \sum_j G_{ij}\langle B_j(t+\tau)B_\ell(t)\rangle \tag{4.8}$$

with exactly the same coefficient matrix G_{ij}.

The advantage of the quantum regression theorem is that it enables the calculation of expressions for two-point correlation functions just from the evolution of single-operator correlation functions.

Let us consider the example of a single electron transistor (SET) at infinite bias ($f_L \to 1$ and $f_R \to 0$). The single-operator expectation values obey

$$\frac{d}{dt}\begin{pmatrix}\langle dd^\dagger(t)\rangle \\ \langle d^\dagger d(t)\rangle\end{pmatrix} = \begin{pmatrix}-\Gamma_L & +\Gamma_R \\ +\Gamma_L & -\Gamma_R\end{pmatrix}\begin{pmatrix}\langle dd^\dagger(t)\rangle \\ \langle d^\dagger d(t)\rangle\end{pmatrix}, \tag{4.9}$$

such that the quantum regression theorem tells us that

$$\frac{d}{d\tau}\begin{pmatrix}\langle dd^\dagger(t+\tau)d^\dagger d(t)\rangle \\ \langle d^\dagger d(t+\tau)d^\dagger d(t)\rangle\end{pmatrix} = \begin{pmatrix}-\Gamma_L & +\Gamma_R \\ +\Gamma_L & -\Gamma_R\end{pmatrix}\begin{pmatrix}\langle dd^\dagger(t+\tau)d^\dagger d(t)\rangle \\ \langle d^\dagger d(t+\tau)d^\dagger d(t)\rangle\end{pmatrix}. \tag{4.10}$$

4.2 Numerical Techniques for Solving Master Equations

Numerical techniques are applicable when analytic methods fail or would require comparably large efforts. We will just discuss two popular approaches here.

4.2.1 Numerical Integration

Numerical integration is generally performed by discretizing time into sufficiently small steps. Note that there are different discretization schemes, e.g., explicit ones,

$$\frac{\rho(t+\Delta t)-\rho(t)}{\Delta t} = \mathscr{L}\rho(t), \tag{4.11}$$

where the right-hand side depends on time t, and implicit ones, such as

$$\frac{\rho(t+\Delta t)-\rho(t)}{\Delta t} = \mathscr{L}\frac{1}{2}[\rho(t)+\rho(t+\Delta t)]. \tag{4.12}$$

Whereas it is straightforward to solve the explicit scheme for $\rho(t+\Delta t)$, in the implicit scheme this would require matrix inversion. Thus, the differential equation is mapped to an iteration equation that maps the density matrix from time t to time $t + \Delta t$. As a rule of thumb, explicit schemes are easy to implement but may be

numerically unstable (i.e., an adaptive stepsize may be required to prevent the solution from exploding) [1]. In contrast, implicit schemes are usually more stable
but require a lot of effort to propagate the solution. Here, we will just discuss a
fourth-order Runge–Kutta solver [2].

In order to propagate a density matrix ρ_n at time t to the density matrix ρ_{n+1} at
time $t + \Delta t$, the fourth-order Runge–Kutta scheme requires the evaluation of four
intermediate values $\sigma_{n,1}, \sigma_{n,2}, \sigma_{n,3}$, and $\sigma_{n,4}$ that can be successively computed from
ρ_n by applying a single multiplication with the Liouvillian $\mathcal{L}$. The density matrix
at time $t + \Delta t$ is then obtained from these auxiliary intermediate values. Explicitly,
the Runge–Kutta algorithm is given by

$$\sigma_{n,1} = \Delta t \mathcal{L} \rho_n,$$

$$\sigma_{n,2} = \Delta t \mathcal{L} \left(\rho_n + \frac{1}{2}\sigma_{n,1} \right),$$

$$\sigma_{n,3} = \Delta t \mathcal{L} \left(\rho_n + \frac{1}{2}\sigma_{n,2} \right), \tag{4.13}$$

$$\sigma_{n,4} = \Delta t \mathcal{L} (\rho_n + \sigma_{n,3}),$$

$$\rho_{n+1} = \rho_n + \frac{1}{6}\sigma_{n,1} + \frac{1}{3}\sigma_{n,2} + \frac{1}{3}\sigma_{n,3} + \frac{1}{6}\sigma_{n,4} + \mathcal{O}\{\Delta t^5\}.$$

This explicit scheme requires four matrix-vector multiplications per time step. It
should always be used in combination with an adaptive stepsize, which can be controlled by comparing (e.g., by computing the norm of the difference) the result from
two successive propagations with stepsize Δt with the result of a single propagation with stepsize $2\Delta t$. If the difference exceeds a predefined error bound, the stepsize must be reduced (and the intermediate result should be discarded). If it is not
too large, the result can always be accepted. If the error estimate is much smaller
than the error bound, one can cautiously increase the time step. In particular when
the matrix-vector multiplication is costly, this will save precious computation time.
Thus, the required computational overhead of 50 % for an adaptive stepsize is well
justified.

Exercise 4.3 (Order of the Runge–Kutta scheme) Acting with the Liouville superoperator performs the time derivative of the density matrix. Show that the presented
scheme (4.13) is of fourth order in Δt, i.e., that

$$\rho_{n+1} = \left[1 + \mathcal{L}\Delta t + \mathcal{L}^2 \frac{\Delta t^2}{2!} + \mathcal{L}^3 \frac{\Delta t^3}{3!} + \mathcal{L}^4 \frac{\Delta t^4}{4!} \right] \rho_n + \mathcal{O}\{\Delta t\}^5.$$

If the Liouvillian $\mathcal{L}$ does not have a special structure, the Runge–Kutta scheme
requires one to store the $N \times N$ density matrix completely. Since N scales exponentially with the size of the system, this may be quite demanding—if not impossible
for larger quantum systems.

4.2.2 Simulation as a Piecewise Deterministic Process (PDP)

Suppose we would like to solve the Lindblad form master equation (in diagonal representation)

$$\dot{\rho} = -\mathrm{i}[H, \rho] + \sum_\alpha \gamma_\alpha \left[L_\alpha \rho L_\alpha^\dagger - \frac{1}{2}\{\rho, L_\alpha^\dagger L_\alpha\} \right] \qquad (4.14)$$

numerically, but we are not able to store the N^2 matrix elements of the density matrix nor to write the master equation in a simpler (e.g., rate equation or block structure) representation.

If it is possible to store at least N states, the master equation can be unraveled to a piecewise deterministic process (PDP) for a pure quantum state. The advantage here lies in the fact that a pure state requires only N complex observables to be evolved.

Consider the nonlinear but deterministic equation

$$|\dot{\Psi}\rangle = -\mathrm{i}\left[H - \frac{\mathrm{i}}{2} \sum_\alpha \gamma_\alpha L_\alpha^\dagger L_\alpha \right] |\Psi\rangle + \frac{1}{2}\left[\sum_\alpha \gamma_\alpha \langle \Psi | L_\alpha^\dagger L_\alpha | \Psi \rangle \right]|\Psi\rangle. \quad (4.15)$$

Although this is nonlinear in $|\Psi(t)\rangle$, one can show that the solution is given by

$$|\Psi\rangle = \frac{e^{-\mathrm{i}Mt}|\Psi_0\rangle}{\langle \Psi_0 | e^{+\mathrm{i}M^\dagger t} e^{-\mathrm{i}Mt} |\Psi_0\rangle^{1/2}}, \qquad (4.16)$$

where we have used the operator $M = H - \frac{\mathrm{i}}{2}\sum_\alpha \gamma_\alpha L_\alpha^\dagger L_\alpha$, which is also often termed the non-hermitian Hamiltonian.

Exercise 4.4 (Norm for continuous evolution) Calculate the norm of the state vector $\langle \Psi(t)|\Psi(t)\rangle$ from Eq. (4.16).

We show the validity of the solution by differentiation

$$|\dot{\Psi}\rangle = -\mathrm{i}M|\Psi\rangle - \frac{1}{2}\frac{e^{-\mathrm{i}Mt}|\Psi_0\rangle}{\langle \Psi_0 | e^{+\mathrm{i}M^\dagger t} e^{-\mathrm{i}Mt}|\Psi_0\rangle^{3/2}}$$

$$\times \mathrm{i}\left[\langle \Psi_0 | e^{+\mathrm{i}M^\dagger t} M^\dagger e^{-\mathrm{i}Mt}|\Psi_0\rangle - \langle \Psi_0 | e^{+\mathrm{i}M^\dagger t} M e^{-\mathrm{i}Mt}|\Psi_0\rangle \right]$$

$$= -\mathrm{i}M|\Psi\rangle - \frac{1}{2}\frac{e^{-\mathrm{i}Mt}|\Psi_0\rangle}{\langle \Psi_0 | e^{+\mathrm{i}M^\dagger t} e^{-\mathrm{i}Mt}|\Psi_0\rangle^{3/2}}\mathrm{i}\langle \Psi_0 | e^{+\mathrm{i}M^\dagger t}\left[M^\dagger - M \right] e^{-\mathrm{i}Mt}|\Psi_0\rangle$$

$$= -\mathrm{i}M|\Psi\rangle + \frac{1}{2}\frac{e^{-\mathrm{i}Mt}|\Psi_0\rangle}{\langle \Psi_0 | e^{+\mathrm{i}M^\dagger t} e^{-\mathrm{i}Mt}|\Psi_0\rangle^{3/2}} \sum_\alpha \gamma_\alpha \langle \Psi_0 | e^{+\mathrm{i}M^\dagger t} L_\alpha^\dagger L_\alpha e^{-\mathrm{i}Mt}|\Psi_0\rangle$$

$$= -\mathrm{i}M|\Psi\rangle + \frac{1}{2}|\Psi\rangle \sum_\alpha \gamma_\alpha \langle \Psi | L_\alpha^\dagger L_\alpha | \Psi \rangle. \qquad (4.17)$$

However, the name PDP already suggests that the process is only piecewise deterministic. To reproduce the original Lindblad dynamics, the continuous evolution (4.16) must be interrupted by stochastic events. The total probability that a jump of the wave function will occur in the infinitesimal interval $[t, t + \Delta t]$ is given by

$$P_{\text{jump}} = \Delta t \sum_\alpha \gamma_\alpha \langle \Psi | L_\alpha^\dagger L_\alpha | \Psi \rangle. \tag{4.18}$$

That is, if a jump has occurred, one still has to decide which jump. Choosing a particular jump

$$|\Psi\rangle \to \frac{L_\alpha |\Psi\rangle}{\sqrt{\langle \Psi | L_\alpha^\dagger L_\alpha | \Psi \rangle}} \tag{4.19}$$

is performed randomly with conditional probability

$$P_\alpha = \frac{\gamma_\alpha \langle \Psi | L_\alpha^\dagger L_\alpha | \Psi \rangle}{\sum_\alpha \gamma_\alpha \langle \Psi | L_\alpha^\dagger L_\alpha | \Psi \rangle}, \tag{4.20}$$

where the normalization is obvious. This recipe for deterministic (continuous) and jump evolutions may also be written as a single stochastic differential equation, which is often called the stochastic Schrödinger equation [3].

Definition 4.2 (Stochastic Schrödinger equation)   A Lindblad-type master equation of the form

$$\dot\rho = -\mathrm{i}[H, \rho] + \sum_\alpha \gamma_\alpha \left[L_\alpha \rho L_\alpha^\dagger - \frac{1}{2}\{\rho, L_\alpha^\dagger L_\alpha\} \right] \tag{4.21}$$

can be effectively modeled by the stochastic differential equation

$$|d\Psi\rangle = \left[-\mathrm{i}H - \frac{1}{2}\sum_\alpha \gamma_\alpha L_\alpha^\dagger L_\alpha + \frac{1}{2}\sum_\alpha \gamma_\alpha \langle \Psi | L_\alpha^\dagger L_\alpha | \Psi \rangle \right] |\Psi\rangle\, dt$$

$$+ \sum_\alpha \left(\frac{L_\alpha |\Psi\rangle}{\sqrt{\langle \Psi | L_\alpha^\dagger L_\alpha | \Psi \rangle}} - |\Psi\rangle \right) dN_\alpha, \tag{4.22}$$

where the Poisson increments dN_α satisfy

$$dN_\alpha\, dN_\beta = \delta_{\alpha\beta}\, dN_\alpha, \qquad \mathscr{E}(dN_\alpha) = \gamma_\alpha \langle \Psi | L_\alpha^\dagger L_\alpha | \Psi \rangle\, dt \tag{4.23}$$

and $\mathscr{E}(x)$ denotes the classical expectation value (ensemble average).

The last two equations simply mean that at most a single jump can occur at once (practically we have $dN_\alpha \in \{0, 1\}$) and that the probability for a jump at time t is

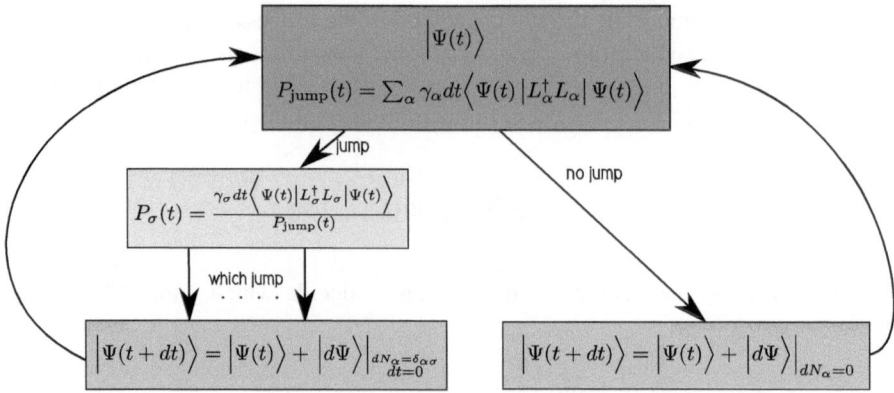

Fig. 4.1 Recipe for propagating the stochastic Schrödinger equation in Definition 4.2. At time t, one calculates the total probability of a jump P_{jump} occurring during the interval $[t, t + \Delta t]$. A random number generator is used to determine whether a jump should occur or not. Given that a jump occurs, one determines which type of jump by drawing another random number: setting the particular $dN_\alpha = 1$ and $dt = 0$, one solves the stochastic Schrödinger equation for $|\Psi(t + \Delta t)\rangle = |\Psi(t)\rangle + |d\Psi(t)\rangle \overset{\alpha}{=} L_\alpha |\Psi(t)\rangle / \sqrt{\langle \Psi(t)|L_\alpha^\dagger L_\alpha|\Psi(t)\rangle}$ and proceeds with the next time step. Given that no jump occurs, one sets $dN_\alpha = 0$ for all α, solves the stochastic Schrödinger equation for $|\Psi(t + \Delta t)\rangle = |\Psi(t)\rangle + |d\Psi(t)\rangle$, and proceeds with the next time step

given by $P_\alpha = \gamma_\alpha \langle \Psi | L_\alpha^\dagger L_\alpha | \Psi \rangle \, dt$. Numerically, it constitutes a simple recipe; see Fig. 4.1. Altogether, the description in terms of a stochastic differential equation in Definition 4.2 simply combines the smooth evolution according to the nonlinear Schrödinger equation (4.15) with stochastic jumps. The concept can be extended beyond Lindblad master equations [4, 5].

It remains to be shown that this PDP is actually an unraveling of the master equation; i.e., that the expectation value of the operator $\hat{\pi} = |\Psi\rangle\langle\Psi|$, also called the covariance matrix,

$$\rho = \mathscr{E}(\hat{\pi}) = \mathscr{E}\big(|\Psi\rangle\langle\Psi|\big), \tag{4.24}$$

fulfills the original Lindblad-type master equation. Then, ensemble averages of all trajectories will also obey the Lindblad dynamics. To show this, we first note that $\langle \Psi | L_\alpha^\dagger L_\alpha | \Psi \rangle = \text{Tr}\{L_\alpha^\dagger L_\alpha \hat{\pi}\}$. The change of the covariance matrix is given by

$$d\hat{\pi} = |d\Psi\rangle\langle\Psi| + |\Psi\rangle\langle d\Psi| + |d\Psi\rangle\langle d\Psi|. \tag{4.25}$$

Note that the last term cannot be neglected completely, since the term $\mathscr{E}(dN_\alpha \, dN_\beta)$ is not necessarily small. Making everything explicit, we obtain

$$d\hat{\pi} = +dt \left\{ -\mathrm{i}[H, \hat{\pi}] - \frac{1}{2} \sum_\alpha \gamma_\alpha \{L_\alpha^\dagger L_\alpha, \hat{\pi}\} + \sum_\alpha \gamma_\alpha \hat{\pi} \, \text{Tr}\{L_\alpha^\dagger L_\alpha \hat{\pi}\} \right\}$$

$$+ \sum_\alpha dN_\alpha \left[\frac{L_\alpha \hat{\pi} L_\alpha^\dagger}{\text{Tr}\{L_\alpha^\dagger L_\alpha \hat{\pi}\}} - \hat{\pi} \right] + \mathscr{O}\{dt^2, dt \, dN_\alpha\}. \tag{4.26}$$

We now use the general relation

$$\mathcal{E}\left(dN_\alpha \, g(\hat{\pi})\right) = \gamma_\alpha \, dt \, \mathcal{E}\left(\text{Tr}\{L_\alpha^\dagger L_\alpha \hat{\pi}\} g(\hat{\pi})\right) \tag{4.27}$$

for arbitrary functions $g(\hat{\pi})$ of the projector. This relation can be understood by binning all K values of the actual state $\hat{\pi}^{(k)}(t)$ in the expectation value into L equal-sized compartments where $\hat{\pi}^{(k)} \approx \hat{\pi}^{(\ell)}$. In each compartment, we have N_ℓ realizations of $dN_\alpha^{\ell m}$ with $1 \leq m \leq N_\ell$ and $\sum_\ell N_\ell = K$, of which we can compute the average first:

$$\mathcal{E}\left(dN_\alpha \, g(\hat{\pi})\right) = \lim_{K \to \infty} \frac{1}{K} \sum_k dN_\alpha^{(k)}(t) \, g\left(\hat{\pi}^{(k)}(t)\right)$$

$$= \lim_{L, N_\ell \to \infty} \frac{\sum_\ell N_\ell \frac{1}{N_\ell} \sum_m dN_\alpha^{(\ell m)} g(\hat{\pi}^{(\ell)}(t))}{\sum_\ell N_\ell}$$

$$= \lim_{L, N_\ell \to \infty} \frac{\sum_\ell N_\ell \gamma_\alpha \, dt \, \text{Tr}\{L_\alpha^\dagger L_\alpha \hat{\pi}^{(\ell)}\} g(\hat{\pi}^{(\ell)}(t))}{\sum_\ell N_\ell}$$

$$= \lim_{L, N_\ell \to \infty} \frac{\sum_\ell N_\ell \gamma_\alpha \, dt \, \mathcal{E}(\text{Tr}\{L_\alpha^\dagger L_\alpha \hat{\pi}^{(\ell)}\} g(\hat{\pi}^{(\ell)}(t)))}{\sum_\ell N_\ell}$$

$$= \gamma_\alpha \, dt \, \mathcal{E}\left(\text{Tr}\{L_\alpha^\dagger L_\alpha \hat{\pi}^{(\ell)}\} g\left(\hat{\pi}^{(\ell)}(t)\right)\right), \tag{4.28}$$

where we have used the relation that $\bar{x} = \frac{\sum_i N_i \bar{x}_i}{\sum_i N_i}$ when $\bar{x}_i$ represent averages of disjoint subsets of the complete set. Specifically, we apply it on the expressions

$$\mathcal{E}\left(dN_\alpha \frac{\hat{\pi}}{\text{Tr}\{L_\alpha^\dagger L_\alpha \hat{\pi}\}}\right) = \gamma_\alpha \, dt \, \mathcal{E}\left(\text{Tr}\{L_\alpha^\dagger L_\alpha \hat{\pi}\} \frac{\hat{\pi}}{\text{Tr}\{L_\alpha^\dagger L_\alpha \hat{\pi}\}}\right) = \gamma_\alpha \, dt \, \rho,$$
$$\mathcal{E}(dN_\alpha \hat{\pi}) = \gamma_\alpha \, dt \, \mathcal{E}\left(\text{Tr}\{L_\alpha^\dagger L_\alpha \hat{\pi}\} \hat{\pi}\right). \tag{4.29}$$

This implies that

$$d\rho = dt \left\{-\text{i}[H, \rho] + \sum_\alpha \gamma_\alpha \left[L_\alpha \rho L_\alpha^\dagger - \frac{1}{2}\{L_\alpha^\dagger L_\alpha, \rho\}\right]\right\}, \tag{4.30}$$

i.e., the average of trajectories from the stochastic Schrödinger equation yields the same solution as the master equation.

This may be of great numerical use: simulating the full master equation for an N-dimensional system Hilbert space may involve the storage of $\mathcal{O}\{N^4\}$ real variables in the Liouvillian, whereas for the generator of the stochastic Schrödinger equation one requires only $\mathcal{O}\{N^2\}$ real variables. This is of course weakened, since in order to get a realistic estimate of the expectation value, one has to compute K different trajectories, but since typically $K \ll N^2$, the stochastic Schrödinger equation is a useful tool in the numeric modeling of a master equation.

4.2.2.1 Example: Cavity in a Thermal Bath

As an example, we study the cavity in a thermal bath. We have the Lindblad-type
master equation describing the interaction of a cavity mode with a thermal bath,

$$\dot{\rho}_S = -i[\Omega a^\dagger a, \rho_S] + \gamma(1 + n_B)\left[a\rho_S a^\dagger - \frac{1}{2}a^\dagger a\rho_S - \frac{1}{2}\rho_S a^\dagger a\right]$$

$$+ \gamma n_B\left[a^\dagger \rho_S a - \frac{1}{2}aa^\dagger \rho_S - \frac{1}{2}\rho_S aa^\dagger\right]. \tag{4.31}$$

We can immediately identify the jump operators

$$L_1 = a \quad \text{and} \quad L_2 = a^\dagger \tag{4.32}$$

and the corresponding rates

$$\gamma_1 = \gamma(1 + n_B) \quad \text{and} \quad \gamma_2 = \gamma n_B. \tag{4.33}$$

From the master equation, we obtain for the occupation number $n = \langle a^\dagger a\rangle$ the evolution equation $\frac{d}{dt}n = -\gamma n + \gamma n_B$, which is solved by

$$n(t) = n_0 e^{-\gamma t} + n_B\left[1 - e^{-\gamma t}\right]. \tag{4.34}$$

The corresponding stochastic differential equation reads

$$|d\Psi\rangle = \left\{-i\Omega a^\dagger a - \frac{1}{2}\left[\gamma(1 + 2n_B)a^\dagger a + \gamma n_B\right]\right.$$

$$+ \frac{1}{2}\left[\gamma(1 + 2n_B)\langle\Psi|a^\dagger a|\Psi\rangle + \gamma n_B\right]\right\}|\Psi\rangle\, dt$$

$$+ \left(\frac{a|\Psi\rangle}{\sqrt{\langle\Psi|a^\dagger a|\Psi\rangle}} - |\Psi\rangle\right) dN_1$$

$$+ \left(\frac{a^\dagger|\Psi\rangle}{\sqrt{\langle\Psi|aa^\dagger|\Psi\rangle}} - |\Psi\rangle\right) dN_2. \tag{4.35}$$

When the initial state is not a superposition of different Fock basis states, the
above equation becomes particularly simple. For example, for a Fock number state
$|\Psi\rangle = |n\rangle$ we obtain

$$|dn\rangle = \left\{-i\Omega n - \frac{1}{2}\left[\gamma(1 + 2n_B)n + \gamma n_B\right] + \frac{1}{2}\left[\gamma(1 + 2n_B)n + \gamma n_B\right]\right\}|n\rangle\, dt$$

$$+ \left(|n - 1\rangle - |n\rangle\right) dN_1 + \left(|n + 1\rangle - |n\rangle\right) dN_2$$

$$= -i\Omega n\, dt\, |n\rangle + \left(|n - 1\rangle - |n\rangle\right) dN_1 + \left(|n + 1\rangle - |n\rangle\right) dN_2 \tag{4.36}$$

such that, provided we start in a single energy eigenstate, superpositions are never created during the evolution. The total probability of having a jump in the system during the interval dt is given by

$$P_{\text{jump}} = \gamma\, dt\left[(1+n_B)\langle n|a^\dagger a|n\rangle + n_B\langle n|aa^\dagger|n\rangle\right]$$
$$= \gamma\left[(1+n_B)n + n_B(n+1)\right]dt. \tag{4.37}$$

If no jump occurs, the system evolves only oscillatory behavior, which has no effect on the expectation value of $a^\dagger a$. However, if a jump occurs, the respective conditional probability of jumping out of the system reads

$$P_1 = \frac{(n_B+1)n}{(n_B+1)n + n_B(n+1)} \tag{4.38}$$

and that of jumping into the system consequently reads (these must add up to one)

$$P_2 = \frac{n_B(n+1)}{(n_B+1)n + n_B(n+1)}. \tag{4.39}$$

Computing trajectories with a suitable random number generator and averaging the trajectories, we find convergence to the master equation result as expected; see Fig. 4.2. The plots in Fig. 4.2 could with the same effort have been obtained by a Monte Carlo solution of the rate equation corresponding to Eq. (4.31),

$$\dot\rho_{nn} = -\gamma\left[n(1+n_B) + (n+1)n_B\right]\rho_{nn} + \gamma(n+1)(1+n_B)\rho_{n+1,n+1}$$
$$+ \gamma n n_B \rho_{n-1,n-1}. \tag{4.40}$$

The rate equation alone however is not sufficient to describe the decay of initial superpositions to a statistical mixture; thus, the stochastic Schrödinger equation is a more general tool.

4.3 Shannon's Entropy Production

We assume that in some favorable basis (e.g., the system energy eigenbasis) the populations of the density matrix $P_i = \rho_{ii}$ obey a rate equation dynamics

$$\dot P_i = \sum_j \mathscr{L}_{ij} P_j = \sum_j \sum_\nu \mathscr{L}_{ij}^{(\nu)} P_j, \tag{4.41}$$

where the rates $\mathscr{L}_{ij}$ from state j to state i are additively decomposable into contributions from different reservoirs ν. Such rate equations are commonly obtained for the quantum optical master equation in Definition 2.4 when nondegenerate system energies are assumed. Furthermore, the assumption of additively entering rates is quite naturally related to the weak coupling limit: it is always possible for an inter-

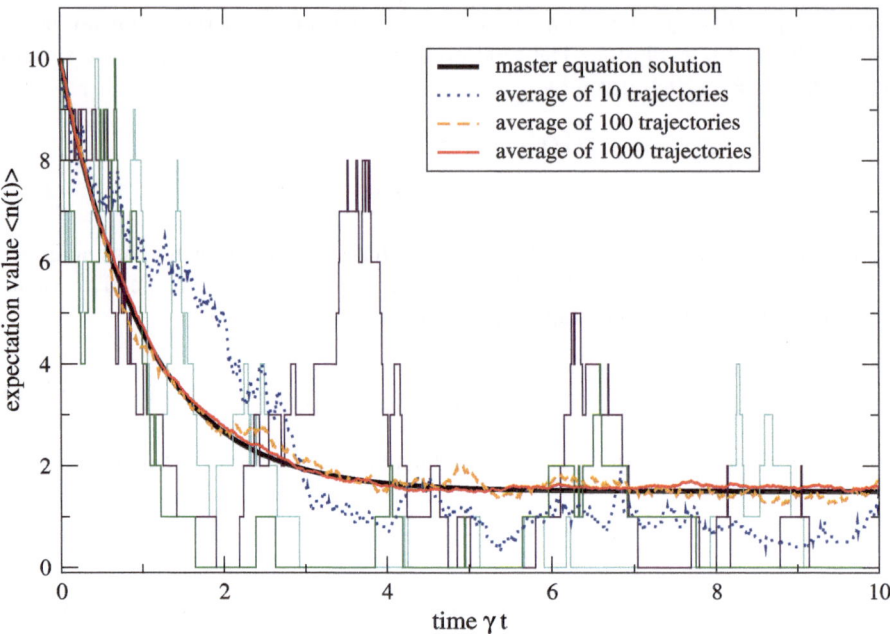

Fig. 4.2 Single trajectories of the stochastic Schrödinger equation (*curves with integer jumps*). The averages of 10, 100, and 1000 trajectories (*thin dotted, dashed,* and *solid curves,* respectively) converge to the prediction from the associated master equation (*thick solid curve*). Parameters have been chosen as $\gamma\, dt = 0.01$, $n_B = 1.5$

action of the form $H_I = \sum_\alpha A_\alpha \otimes B_\alpha$ with system and bath operators A_α and B_α, respectively, to choose $\langle B_\alpha \rangle = 0$. For L multiple reservoirs kept at different equilibrium states, the stationary density matrix is given by a tensor product of different equilibrium states

$$\bar{\rho} = \frac{e^{-\beta_1(H_B^{(1)} - \mu_1 N_B^{(1)})}}{Z_1} \otimes \cdots \otimes \frac{e^{-\beta_L(H_B^{(L)} - \mu_L N_B^{(L)})}}{Z_L}, \qquad (4.42)$$

where β_ν and μ_ν represent the temperature and chemical potential of the ν-th reservoir described by the Hamiltonian $H_B^{(\nu)}$ and with total particle number operator $N_B^{(\nu)}$.

Exercise 4.5 (Additivity of rates) Show that for an interaction Hamiltonian of the form $H_I = \sum_\alpha A_\alpha \otimes B_\alpha = \sum_a \sum_\nu A_{a\nu} \otimes B_{a\nu}$, where ν labels the reservoir and where $\langle B_{a\nu} \rangle = 0$ holds, different reservoirs do not interfere, such that the rates can be calculated additively:

$$C_{\alpha\beta}(\tau) = C_{a\nu,b\mu}(\tau) = \langle \boldsymbol{B}_{a\nu}(\tau) B_{b\nu} \rangle \delta_{\mu\nu}.$$

4.3.1 Balance Equation Far from Equilibrium

Keeping in mind that each reservoir is kept at a certain equilibrium, we postulate the existence of a local detailed balance condition for each reservoir. This implies that the ratio of forward and backward transition rates between states i and j that are triggered by reservoir ν obey

$$\frac{\mathscr{L}_{ji}^{(\nu)}}{\mathscr{L}_{ij}^{(\nu)}} = e^{-\beta_\nu[(\varepsilon_j - \varepsilon_i) - \mu_\nu(n_j - n_i)]}, \tag{4.43}$$

where β_ν and μ_ν denote the inverse temperature and chemical potential of the corresponding reservoir, and ε_i and n_i denote the energy and particle number of the state i, respectively. The above relation follows naturally from the extension of the Kubo–Martin–Schwinger (KMS) condition (2.51) to systems with chemical potentials and is automatically fulfilled for a large number of microscopically derived models, as we shall see later.

Then, the Shannon entropy of the system, $S = -\sum_i P_i(t) \ln P_i(t)$, obeys a balance equation,

$$\dot{S} = -\frac{d}{dt} \sum_i P_i \ln P_i = -\sum_i \dot{P}_i \ln P_i$$

$$= -\sum_{ij} \sum_\nu \mathscr{L}_{ij}^{(\nu)} P_j \ln\left(P_i \frac{\mathscr{L}_{ji}^{(\nu)}}{P_j \mathscr{L}_{ij}^{(\nu)}} \frac{P_j \mathscr{L}_{ij}^{(\nu)}}{\mathscr{L}_{ji}^{(\nu)}} \right)$$

$$= +\sum_{ij} \sum_\nu \mathscr{L}_{ij}^{(\nu)} P_j \ln\left(\frac{\mathscr{L}_{ij}^{(\nu)} P_j}{\mathscr{L}_{ji}^{(\nu)} P_i} \right) + \sum_{ij} \sum_\nu \mathscr{L}_{ij}^{(\nu)} P_j \ln\left(\frac{\mathscr{L}_{ji}^{(\nu)}}{\mathscr{L}_{ij}^{(\nu)}} \frac{1}{P_j} \right)$$

$$= +\sum_{ij} \sum_\nu \mathscr{L}_{ij}^{(\nu)} P_j \ln\left(\frac{\mathscr{L}_{ij}^{(\nu)} P_j}{\mathscr{L}_{ji}^{(\nu)} P_i} \right) + \sum_{ij} \sum_\nu \mathscr{L}_{ij}^{(\nu)} P_j \ln\left(\frac{\mathscr{L}_{ji}^{(\nu)}}{\mathscr{L}_{ij}^{(\nu)}} \right)$$

$$\overset{t\to\infty}{\to} + \underbrace{\sum_{ij} \sum_\nu \mathscr{L}_{ij}^{(\nu)} \bar{P}_j \ln\left(\frac{\mathscr{L}_{ij}^{(\nu)} \bar{P}_j}{\mathscr{L}_{ji}^{(\nu)} \bar{P}_i} \right)}_{\geq 0} + \sum_{ij} \sum_\nu \mathscr{L}_{ij}^{(\nu)} \bar{P}_j \underbrace{\ln\left(\frac{\mathscr{L}_{ji}^{(\nu)}}{\mathscr{L}_{ij}^{(\nu)}} \right)}_{-\beta_\nu[(\varepsilon_j - \varepsilon_i) - \mu_\nu(n_j - n_i)]} \; .$$

$$\tag{4.44}$$

In the above lines, we have simply used trace conservation $\sum_i \mathscr{L}_{ij}^{(\nu)} = 0$ and finally the local detailed balance property (4.43). This property enables us to identify in the long-term limit the second term as energy and matter currents. When multiplied by the inverse temperature of the corresponding reservoir, they combine to the entropy change rate of the reservoirs, which motivates the definition below.

Definition 4.3 (Entropy flow)   For a rate equation of the type (4.41), the entropy flow from reservoir v is defined as

$$\dot{S}_{\mathrm{e}}^{(v)} = +\sum_{ij} \mathscr{L}_{ij}^{(v)} \bar{P}_j \left[-\beta_v \left[(\varepsilon_j - \varepsilon_i) - \mu_v (n_j - n_i) \right] \right]$$

$$= \beta_v \left(I_E^{(v)} - \mu_v I_M^{(v)} \right), \tag{4.45}$$

where energy currents $I_E^{(v)}$ and matter currents $I_M^{(v)}$ associated to reservoir v count positive when entering the system.

The remaining contribution corresponds to a production term [6]. We note that it is always positive, which can be deduced from the formal similarity to the Kullback–Leibler divergence of two probability distributions or, more directly, by using the logarithmic sum inequality.

Exercise 4.6 (Logarithmic sum inequality)   Show that for non-negative a_i and b_i,

$$\sum_{i=1}^{n} a_i \ln \frac{a_i}{b_i} \geq a \ln \frac{a}{b}$$

with $a = \sum_i a_i$ and $b = \sum_i b_i$.

Its positivity is perfectly consistent with the second law of thermodynamics, and we therefore identify the remaining contribution as entropy production.

Definition 4.4 (Entropy production)   For a rate equation of the type (4.41), the average entropy production is defined as

$$\dot{S}_{\mathrm{i}} = \sum_{ij} \sum_{v} \mathscr{L}_{ij}^{(v)} \bar{P}_j \ln \left(\frac{\mathscr{L}_{ij}^{(v)} \bar{P}_j}{\mathscr{L}_{ji}^{(v)} \bar{P}_i} \right) \geq 0. \tag{4.46}$$

It is always positive and at steady state balanced by the entropy flow.

When the dimension of the system's Hilbert space is finite and the rate equation (4.41) approaches a stationary state, its Shannon entropy will also approach a constant value $\dot{S} = 0$. Therefore, at steady state the entropy production in the system must be balanced by the entropy flow through its terminals

$$\dot{S}_{\mathrm{i}} = -\dot{S}_{\mathrm{e}} = -\sum_v \beta_v \left(I_E^{(v)} - \mu_v I_M^{(v)} \right). \tag{4.47}$$

The above formula conveniently relates the entropy production to energy and matter currents from the terminals into the system. Evidently, the entropy production is thus related to heat currents $\dot{Q}^{(v)} = I_E^{(v)} - \mu_v I_M^{(v)}$, which can be determined from a master equation by means of the full counting statistics.

We note here that identifying the entropy production in a system is not a purely academic exercise: in the long term, it is additive in the respective entropy flows, and their identification allows, e.g., for the definition of thermodynamically meaningful (and bounded) efficiencies of thermoelectric nanoscale devices.

4.3.2 Linear Response for Two Terminals

As an example, we consider a system coupled to two terminals S and D obeying a rate equation dynamics as discussed before. In the long-time limit, entropy production will be balanced by the entropy flow, and assuming that both energy and matter currents are conserved, $I_E^{(D)} + I_E^{(S)} = 0$ and $I_M^{(D)} + I_M^{(S)} = 0$, we can express the entropy production solely using the currents entering the system from the source $\dot{S}_i = (\beta_D - \beta_S)I_E^{(S)} + (\mu_S\beta_S - \mu_D\beta_D)I_M^{(S)}$. In the linear response regime we assume that the differences of temperatures and chemical potentials are small. Rewriting these parameters in terms of mean and differences $\beta_S = \beta - \Delta\beta/2$, $\beta_D = \beta + \Delta\beta/2$, $\mu_S = \mu + \Delta\mu/2$, and $\mu_D = \mu - \Delta\mu/2$, the entropy production can be expanded in $\Delta\beta$ and $\Delta\mu$, which to lowest order yields

$$\dot{S}_i = \Delta\beta\left(I_E^{(S)} - \mu I_M^{(S)}\right) + \beta\Delta\mu I_M^{(S)} = \Delta\beta\dot{Q} + \beta\Delta\mu I_M, \qquad (4.48)$$

where $\dot{Q}$ represents the heat current and I_M the matter current from S to D, respectively. This equation has the characteristic affinity-flux form [7], where the affinity to the heat current is given by $\Delta\beta = \beta_D - \beta_S = \Delta T/T^2 + \mathcal{O}\{\Delta T^2\}$, and the affinity for the matter current is given by $\beta\Delta\mu = \beta(\mu_S - \mu_D)$. In the linear response regime, the fluxes are linearly related to the affinities,

$$\begin{pmatrix} \dot{Q} \\ I_M \end{pmatrix} = \begin{pmatrix} L_{QQ} & L_{QM} \\ L_{MQ} & L_{MM} \end{pmatrix} \begin{pmatrix} \Delta\beta \\ \beta\Delta\mu \end{pmatrix}, \qquad (4.49)$$

with the Onsager matrix L. Consequently, the entropy production can—in the linear response—be expressed as a quadratic form of the affinities

$$\dot{S}_i = (\Delta\beta, \beta\Delta\mu) \begin{pmatrix} L_{QQ} & L_{QM} \\ L_{MQ} & L_{MM} \end{pmatrix} \begin{pmatrix} \Delta\beta \\ \beta\Delta\mu \end{pmatrix}. \qquad (4.50)$$

Positivity of the entropy production requires positivity of the Onsager matrix.

Considering, e.g., an SET with the matter current in the weak coupling regime approaching Eq. (3.50), and assuming tight coupling between energy and matter currents, such that $\dot{Q} = (\varepsilon - \mu)I_M$ (compare also Sect. 5.1), the Onsager relations become

$$\begin{pmatrix} \dot{Q} \\ I_M \end{pmatrix} = \frac{\Gamma_S\Gamma_D}{\Gamma_S + \Gamma_D} f(1 - f) \begin{pmatrix} (\varepsilon - \mu)^2 & (\varepsilon - \mu) \\ (\varepsilon - \mu) & 1 \end{pmatrix} \begin{pmatrix} \Delta\beta \\ \beta\Delta\mu \end{pmatrix}$$

$$\text{with } f = \frac{1}{e^{\beta(\varepsilon - \mu)} + 1}, \qquad (4.51)$$

which fulfills the Onsager relation $L_{QM} = L_{MQ}$ and has a positive definite Onsager matrix. Due to the tight coupling property we note that the determinant of the Onsager matrix vanishes.

Exercise 4.7 (SET Onsager relations) Confirm the validity of Eq. (4.51).

4.4 Full Counting Statistics: Phenomenological Introduction

Having successfully derived a rate equation of the form (4.41), one can very often interpret the process associated with the rate $\mathscr{L}_{ij}^{(\nu)}$ as a jump of ($|n_i - n_j|$) particles from the bath ν into the system (when $n_i > n_j$) or out of the system into the bath ν (when $n_i < n_j$). Typically, the weak coupling limit assumed during the derivation of the rate equation leads to sequential transport only; i.e., only terms $\mathscr{L}_{ij}^{(\nu)}$ with $n_i - n_j \in \{-1, 0, +1\}$ will be nonvanishing. Such a jump may also transfer the energy $|E_i - E_j|$ from the bath ν into the system ($E_i > E_j$) or out of the system into the bath ν ($E_i < E_j$), even if no particles are transferred ($n_i = n_j$). A straightforward observation is that even though on average a matter or energy current may be directed in a certain direction, there is for a given initial state a finite probability that a jump will occur in the opposite direction. Such trajectories would actually decrease the entropy of the system and must—since they are not completely forbidden—somehow be suppressed to obey the second law on average. Fortunately, one may calculate the statistics of these events in a straightforward manner, as will be discussed in the following subsections.

4.4.1 Discrete Particle Counting Statistics

We denote the probability that the system is in the state i and simultaneously n particles have tunneled into reservoir σ by $P_i^{(n)}(t)$. Obviously, we have $-\infty < n < +\infty$ (unless transport is unidirectional) and $P_i(t) = \sum_n P_i^{(n)}(t)$. However, the rate equation (4.41) can now be written as

$$\dot{P}_i^{(n)} = \sum_{\nu \neq \sigma} \sum_j \mathscr{L}_{ij}^{(\nu)} P_j^{(n)} + \mathscr{L}_{ii}^{(\sigma)} P_i^{(n)} + \sum_{j \neq i} \mathscr{L}_{ij}^{(\sigma)} P_j^{(n+n_i-n_j)}, \qquad (4.52)$$

where we have separated the jumps triggered by other reservoirs than σ and also the trace-preserving diagonal term proportional to $\mathscr{L}_{ii}^{(\sigma)}$. We note that one can interpret the term $\mathscr{L}_{ij}^{(\sigma)} P_j^{(n+n_i-n_j)}$ as follows: before the jump, the system is in state j with n_j particles in the system and $n + n_i - n_j$ particles in reservoir σ. After the jump, the system is in state i with n_i particles in the system and n particles in reservoir σ. Thus, the combined particle number $n + n_i$ in both system and reservoir is conserved during the jump.

For ease of notation, we write the conditioned rate equation (4.52) as a conditioned density vector $\rho^{(n)} = (P_1^{(n)}, \ldots, P_d^{(n)})^\mathsf{T}$ and assume that at most one particle can be transferred at once to and from the bath. This is the standard case arising in most microscopic derivations; however, for a counter-example we refer the reader to Sect. 5.8. Then, Eq. (4.52) becomes

$$\dot{\rho}^{(n)} = \mathscr{L}_0 \rho^{(n)} + \mathscr{L}_- \rho^{(n+1)} + \mathscr{L}_+ \rho^{(n-1)}, \tag{4.53}$$

and the translational invariance in n (the rates contained in $\mathscr{L}_{0/\pm}$ do not depend on n themselves) suggests that we simplify the coupled system via a discrete Fourier transformation,

$$\rho(\chi, t) = \sum_n \rho^{(n)}(t) e^{in\chi}, \tag{4.54}$$

which yields a d-dimensional ordinary differential equation similar to a rate equation but now with complex-valued rates, since we have introduced the counting field χ:

$$\dot{\rho}(\chi, t) = \left[\mathscr{L}_0 + e^{-i\chi} \mathscr{L}_- + e^{+i\chi} \mathscr{L}_+ \right] \rho(\chi, t) = \mathscr{L}(\chi) \rho(\chi, t). \tag{4.55}$$

Thus, we have reduced the dimension at the price of introducing a dimensionless counting field, but the resulting generalized master equation can now be formally solved as

$$\rho(\chi, t) = e^{\mathscr{L}(\chi)t} \rho(\chi, 0) = e^{\mathscr{L}(\chi)t} \rho_0, \tag{4.56}$$

where we have used the convention that at time $t = 0$ no particles should have entered the reservoir $\rho^{(n)}(0) = \rho_0 \delta_{n,0}$.

If we disregard the state of the system and only consider the number of tunneled particles, the corresponding probability becomes

$$P_n(t) = \sum_i P_i^{(n)}(t) = \mathrm{Tr}\{\rho^{(n)}(t)\} = \frac{1}{2\pi} \int_{-\pi}^{+\pi} \mathrm{Tr}\{e^{\mathscr{L}(\chi)t} \rho_0\} e^{-in\chi} d\chi, \tag{4.57}$$

where we have simply inserted the inverse of the discrete Fourier transform (4.54). By tracing over Eq. (4.54) and taking suitable derivatives with respect to the counting field χ, we note that the moments of this probability distribution can be conveniently calculated by taking derivatives:

$$\langle n^k \rangle \equiv \sum_n n^k P_n(t) = (-i\partial_\chi)^k \mathrm{Tr}\{\rho(\chi, t)\}\big|_{\chi=0}$$

$$= (-i\partial_\chi)^k \mathrm{Tr}\{e^{\mathscr{L}(\chi)t} \rho_0\}\big|_{\chi=0}. \tag{4.58}$$

This directly motivates us to define a moment-generating function.

Definition 4.5 (Moment-generating function)   With a particle-counting-field dependent Liouvillian $\mathscr{L}(\chi)$, the moment-generating function corresponding to the distribution $P_n(t)$ is defined as

$$M(\chi,t) = \mathrm{Tr}\left\{e^{\mathscr{L}(\chi)t}\rho_0\right\} \stackrel{t\to\infty}{\to} \mathrm{Tr}\left\{e^{\mathscr{L}(\chi)t}\bar{\rho}\right\} \qquad (4.59)$$

with the initial state ρ_0 and the stationary state defined by $\mathscr{L}(0)\bar{\rho} = \mathbf{0}$.

Given the moment-generating function, moments of the distribution $P_n(t)$ may be calculated conveniently via

$$\langle n^k \rangle(t) = (-i\partial_\chi)^k M(\chi,t)\big|_{\chi=0}, \qquad (4.60)$$

whereas the calculation of the full distribution requires one to calculate the full inverse Fourier transform of the moment-generating function,

$$P_n(t) = \frac{1}{2\pi} \int_{-\pi}^{+\pi} M(\chi,t)e^{-in\chi}\,d\chi. \qquad (4.61)$$

The latter is, except for some specific cases, only numerically possible.

4.4.2   Continuous Energy Counting Statistics

Similarly, we may extract the statistics of energy transfers from the rate equation (4.41). One possible way [8] is to treat transitions occurring with a certain energy transfer ω_i with a separate particle number counting n_i and a separate dimensionless counting field χ_i. The total transferred energy can then later be deduced from the specific particle transitions via $E = \sum_i \omega_i n_i$. In this case, the energy-resolved distribution function would then be given by

$$\rho^{(E)}(t) = \sum_{n_1,\dots,n_k} \rho^{(n_1\cdots n_k)}(t)\delta\left(E - \sum_i \delta E_i n_i\right). \qquad (4.62)$$

Here however, we would like avoid introducing too many counting fields and therefore decide to count the transferred energy directly [9]. Obviously, when the transition frequencies of the system ω_i are incommensurate, the total transferred energy E will become a continuous variable.

Denoting the density vector conditioned on energy E contained in the reservoir σ by $\rho^{(E)} = (P_1^{(E)}, \dots, P_d^{(E)})^{\mathrm{T}}$ with $-\infty < E/\Omega < \infty$ (for any energy scale Ω) and $\rho(t) = \int \rho^{(E)}(t)\,dE$, we may write the rate equation (4.41) as

$$\dot{\rho}^{(E)} = \mathscr{L}_0\rho^{(E)} + \sum_{\Delta E} \mathscr{L}_{\Delta E}\rho^{(E-\Delta E)}, \qquad (4.63)$$

where $\mathscr{L}_0$ does not induce energy transfers with reservoir σ and $\mathscr{L}_{\Delta E}$ describes the transfer of energy ΔE from the system to reservoir σ; negative ΔE simply implies the opposite direction. Here, one usually has multiple energy differences $|\Delta E|$. Only very simple systems admit only a single transition frequency, and then energy and particle currents are tightly coupled. Now, we have to choose a continuous Fourier transform

$$\rho(\xi, t) = \int \rho^{(E)}(t) e^{iE\xi} \, dE, \tag{4.64}$$

where the dual field ξ now has the dimension of inverse energy. The Fourier-transformed master equation becomes

$$\dot{\rho}(\xi, t) = \left[\mathscr{L}_0 + \sum_{\Delta E} \mathscr{L}_{\Delta E} e^{i\xi \Delta E} \right] \rho(\xi, t) = \mathscr{L}(\xi) \rho(\xi, t), \tag{4.65}$$

and the field ξ is now allowed to range over the complete real axis. With the convention that initially no energy has been transferred, $\rho^{(E)}(0) = \delta(E)\rho_0$, we may similarly write the solution as $\rho(\xi, t) = e^{\mathscr{L}(\xi)t} \rho_0$.

The moments of the energy-transfer distribution

$$\langle E^k \rangle = \int E^k \operatorname{Tr}\{\rho^{(E)}(t)\} \, dE \tag{4.66}$$

can now be similarly obtained—compare Eq. (4.64)—by differentiation of the moment-generating function

$$M(\xi, t) = \operatorname{Tr}\{e^{\mathscr{L}(\xi)t} \rho_0\} \tag{4.67}$$

with respect to the dimensioned counting field ξ. Similarly, the full distribution can be obtained by calculating the inverse Fourier transform of the moment-generating function

$$P_E(t) = \frac{1}{2\pi} \int_{-\infty}^{+\infty} M(\xi, t) e^{-in\xi} \, d\xi. \tag{4.68}$$

4.4.3 Moments and Cumulants

It is often more convenient to characterize distributions by cumulants instead of moments, since higher cumulants are invariant against translations of the distribution (in the following discussion we treat dimensionless particle counting and dimensioned energy counting similarly).

Definition 4.6 (Cumulant-generating function) The cumulant-generating function is defined as the logarithm of the moment-generating function:

$$C(\chi, t) = \ln \operatorname{Tr}\{e^{\mathscr{L}(\chi)t} \rho_0\}. \tag{4.69}$$

The cumulants of the distribution $P_n(t)$ are obtained by differentiation with respect to the counting field

$$\langle\langle n^k \rangle\rangle(t) = (-i\partial_\chi)^k C(\chi,t)\big|_{\chi=0}, \tag{4.70}$$

and similarly for cumulants of the energy distribution function. Cumulants and moments are therefore obviously related. Considering for example particle counting, the first few cumulants can be expressed by the moments as

$$
\begin{aligned}
\langle\langle n^1 \rangle\rangle &= \langle n^1 \rangle, \\
\langle\langle n^2 \rangle\rangle &= \langle n^2 \rangle - \langle n \rangle^2, \\
\langle\langle n^3 \rangle\rangle &= \langle n^3 \rangle - 3\langle n\rangle\langle n^2\rangle + 2\langle n\rangle^3, \\
\langle\langle n^4 \rangle\rangle &= \langle n^4 \rangle - 4\langle n\rangle\langle n^3\rangle - 3\langle n^2\rangle^2 + 12\langle n\rangle^2\langle n^2\rangle - 6\langle n\rangle^4,
\end{aligned}
\tag{4.71}
$$

and they geometrically correspond to the mean, width, skewness, and kurtosis of a distribution, respectively. It should be noted however that such simple geometric interpretations only hold for unimodal distributions.

The true advantage of considering cumulants instead of moments becomes visible for master equations admitting only a single stationary state. Then, the cumulant-generating function in the large-time limit scales approximately linearly in time,

$$C(\chi,t) \to \lambda(\chi)t, \tag{4.72}$$

where $\lambda(\chi)$ is the eigenvalue of the generalized Liouvillian $\mathscr{L}(\chi)$ with the largest real part.

We show this by using the decomposition of the Liouvillian in Jordan block form,

$$\mathscr{L}(\chi) = Q(\chi)\mathscr{L}_J(\chi)Q^{-1}(\chi), \tag{4.73}$$

where $Q(\chi)$ is a (in general nonunitary) similarity matrix and $\mathscr{L}_J(\chi)$ contains the eigenvalues of the Liouvillian on its diagonal, distributed in blocks with a size corresponding to the eigenvalue multiplicity. We assume that there exists only one stationary state $\bar\rho$, i.e., only one eigenvalue $\lambda(\chi)$ with $\lambda(0) = 0$, and that all other eigenvalues have a nonvanishing negative real part near $\chi = 0$. Then, we use this decomposition in the matrix exponential to estimate its long-term evolution:

$$
\begin{aligned}
\mathscr{M}(\chi,t) &= \mathrm{Tr}\{e^{\mathscr{L}(\chi)t}\rho_0\} = \mathrm{Tr}\{e^{Q(\chi)\mathscr{L}_J(\chi)Q^{-1}(\chi)t}\rho_0\} \\
&= \mathrm{Tr}\{Q(\chi)e^{\mathscr{L}_J(\chi)t}Q^{-1}(\chi)\rho_0\} \\
&\to \mathrm{Tr}\left\{Q(\chi)\begin{pmatrix} e^{\lambda(\chi)\cdot t} & & \\ & 0 & \\ & & \ddots \\ & & & 0 \end{pmatrix}Q^{-1}(\chi)\rho_0\right\}
\end{aligned}
$$

$$= e^{\lambda(\chi)\cdot t} \operatorname{Tr}\left\{ Q(\chi) \begin{pmatrix} 1 & & & \\ & 0 & & \\ & & \ddots & \\ & & & 0 \end{pmatrix} Q^{-1}(\chi)\rho_0 \right\}$$

$$= e^{\lambda(\chi)t} c(\chi) \tag{4.74}$$

with some polynomial $c(\chi)$ depending on the matrix $Q(\chi)$ and on the initial state ρ_0. This implies that the cumulant-generating function

$$\mathscr{C}(\chi, t) = \ln \mathscr{M}(\chi, t) = \lambda(\chi)t + \ln c(\chi) \approx \lambda(\chi)t \tag{4.75}$$

becomes linear in $\lambda(\chi)$ for large times, up to a small correction. This small correction is usually negligible, particularly when one is interested in time derivatives such as the current. We note here that this simple limit only holds when there is a unique stationary state. For bistable or multistable systems a more sophisticated theory applies [10, 11]. Note further, that when cumulants are to be obtained from the moments, the small constant correction may be important.

Exercise 4.8 (Cumulant-generating function) Calculate the long-term cumulant-generating function for current through the SET

$$\mathscr{L}(\chi) = \begin{pmatrix} -\Gamma_L f_L - \Gamma_R f_R & +\Gamma_L(1 - f_L) + \Gamma_R(1 - f_R)e^{+i\chi} \\ +\Gamma_L f_L + \Gamma_R f_R e^{-i\chi} & -\Gamma_L(1 - f_L) - \Gamma_R(1 - f_R) \end{pmatrix}.$$

What are the first two long-term cumulants for the current, i.e., current $I = \frac{d}{dt}\langle\langle n \rangle\rangle$ and noise $S = \frac{d}{dt}\langle\langle n^2 \rangle\rangle = \frac{d}{dt}(\langle n^2 \rangle - \langle n \rangle^2)$?

4.4.4 Convenient Calculation of Lower Cumulants

To calculate moments and/or cumulants, it is not always necessary to exponentiate the Liouvillian or to calculate its dominant eigenvalue.

If one is just interested in the long-term current, e.g., the time derivative of the mean energy or particle number transferred (first moment/cumulant), the calculations are considerably simplified, since we can insert the stationary state as initial condition:

$$I = \langle \dot{n}(t) \rangle = -i\partial_\chi \frac{d}{dt} \operatorname{Tr}\{e^{\mathscr{L}(\chi)t}\bar{\rho}\}\Big|_{\chi=0} = -i\partial_\chi \operatorname{Tr}\{\mathscr{L}(\chi)e^{\mathscr{L}(\chi)t}\bar{\rho}\}\Big|_{\chi=0}$$

$$= -i\operatorname{Tr}\{\mathscr{L}'(0)e^{\mathscr{L}(0)t}\bar{\rho}\} - i\operatorname{Tr}\{\mathscr{L}(0)\partial_\chi e^{\mathscr{L}(\chi)t}\big|_{\chi=0}\bar{\rho}\}$$

$$= -i\operatorname{Tr}\{\mathscr{L}'(0)\bar{\rho}\}, \tag{4.76}$$

where we have used $\mathscr{L}(0)\bar{\rho} = 0$ and also $\text{Tr}\{\mathscr{L}(0)S\} = 0$ for all operators S (trace conservation). Therefore, to compute the current, the only challenge is to calculate the stationary state of the rate matrix at vanishing counting fields.

To calculate the long-term limit of higher cumulants, we may also use limits on the Laplace-transformed moment-generating function:

$$\tilde{M}(\chi, z) = \int_0^\infty M(\chi, t) e^{-zt} \, dt = \text{Tr}\left\{ \frac{1}{z\mathbf{1} - \mathscr{L}(\chi)} \rho_0 \right\}$$

$$\rightarrow \text{Tr}\left\{ \frac{1}{z\mathbf{1} - \mathscr{L}(\chi)} \bar{\rho} \right\}. \tag{4.77}$$

Having calculated the first moment $\langle n \rangle = It$, the time derivative of the second cumulant is, e.g., related to the first two moments via

$$C_2 = \lim_{t \to \infty} \frac{d}{dt} \left[\langle n^2 \rangle - \langle n \rangle^2 \right] = \lim_{t \to \infty} \left[\frac{d}{dt} \langle n^2 \rangle - 2I^2 t \right]. \tag{4.78}$$

Performing a Laplace transform of this equation, we may use well-known properties of this transform,

$$f(t) \leftrightarrow \tilde{f}(z) \quad \Longrightarrow \quad \dot{f}(t) \leftrightarrow z\tilde{f}(z) - f(0),$$

$$\lim_{t \to \infty} f(t) = \lim_{z \to 0} z\tilde{f}(z), \tag{4.79}$$

to obtain an alternative formula for the time derivative of the second cumulant:

$$C_2 = \lim_{z \to 0} z \left[z(-i\partial_\chi)^2 \tilde{M}(\chi, z) - \frac{2I^2}{z^2} \right], \tag{4.80}$$

where we have used the fact that the initial value of the second moment vanishes. The evaluation of this expression requires only knowledge of the stationary current I and the inverse matrix occurring in the Laplace transform of the moment-generating function—which is much simpler to calculate than a matrix exponential. Keeping in mind that cumulants may have a constant contribution, one may extend the scheme to obtain formulae for higher cumulants.

4.4.5 Fluctuation Theorems

Representing the full energy or particle distributions, not in terms of the moment-generating function in Eqs. (4.61) and (4.68), but with the cumulant-generating

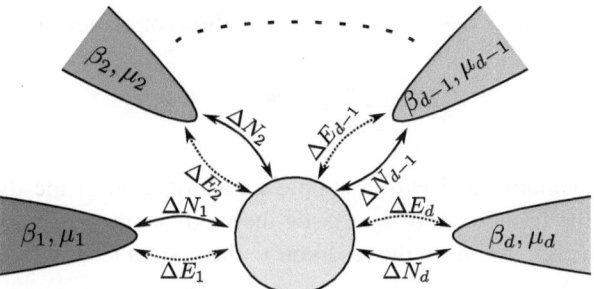

Fig. 4.3 Sketch of a system (*yellow circle*) that is coupled to d terminals, which admits the exchange of matter ΔN_i and energy ΔE_i between system and reservoirs. When these are in thermal equilibrium states described by temperatures and chemical potentials, one finds for sufficiently weak couplings a fluctuation theorem of energy and matter exchanges (Color figure online)

function $\mathscr{C}(\chi/\xi, t)$, we obtain

$$
P_{\Delta n}(t) = \frac{1}{2\pi} \int_{-\pi}^{+\pi} e^{\mathscr{C}(\chi,t)} e^{-i\Delta n \chi}\, d\chi,
$$
$$
P_{\Delta E}(t) = \frac{1}{2\pi} \int_{-\infty}^{+\infty} e^{\mathscr{C}(\xi,t)} e^{-i\Delta E \xi}\, d\xi.
\tag{4.81}
$$

We will consider the general case here where all matter and energy transfers are monitored for a system with d junctions; see Fig. 4.3. Formally, simultaneous counting at all junctions requires introducing the multidimensional vectors $\Delta \boldsymbol{n} = (\Delta n_1, \ldots, \Delta n_d)$ and $\Delta \boldsymbol{E} = (\Delta E_1, \ldots, \Delta E_d)$, where Δn_ν and ΔE_ν denote the particles and energy exchanged with the νth reservoir and the system (counted positive by construction when entering the system), respectively. The corresponding probability distribution reads

$$
P_{+\Delta \boldsymbol{n}, +\Delta \boldsymbol{E}}(t) = \left(\frac{1}{2\pi}\right)^{2d} \int_{-\pi}^{+\pi} \cdots \int_{-\pi}^{+\pi} d^d\chi \int_{-\infty}^{+\infty} \cdots \int_{-\infty}^{+\infty} d^d\xi
$$
$$
\times\, e^{\mathscr{C}(\chi,\xi,t)}\, e^{-i\Delta \boldsymbol{n}\cdot\chi}\, e^{-i\Delta \boldsymbol{E}\cdot\xi},
\tag{4.82}
$$

such that the probability of the inverse process is

$$
P_{-\Delta \boldsymbol{n}, -\Delta \boldsymbol{E}}(t) = \left(\frac{1}{2\pi}\right)^{2d} \int_{-\pi}^{+\pi} \cdots \int_{-\pi}^{+\pi} d^d\chi \int_{-\infty}^{+\infty} \cdots \int_{-\infty}^{+\infty} d^d\xi
$$
$$
\times\, e^{\mathscr{C}(-\chi,-\xi,t)}\, e^{-i\Delta \boldsymbol{n}\cdot\chi}\, e^{-i\Delta \boldsymbol{E}\cdot\xi},
\tag{4.83}
$$

where we have already transformed the integration variables $\chi \to -\chi$ and $\xi \to -\xi$. When now the cumulant-generating function obeys a symmetry of the form (typically, such symmetries arise in the long-term limit)

$$
\mathscr{C}(-\chi, -\xi, t) = \mathscr{C}(\chi + i\Delta_\chi, \xi + i\Delta_\xi, t),
\tag{4.84}
$$

this implies a fluctuation theorem for the probabilities of matter and energy transfers

$$\frac{P_{+\Delta n,+\Delta E}}{P_{-\Delta n,-\Delta E}} = e^{\Delta n \cdot \Delta_\chi} e^{\Delta E \cdot \Delta_\xi}, \tag{4.85}$$

which can be demonstrated with a simple transformation of the integrand. Interpreted within the framework of stochastic thermodynamics [12], a transfer of Δn particles and energy ΔE from the reservoir to the system leads to the production of entropy of

$$\Delta_i S = \sum_{\nu=1}^{d} \beta_\nu \mu_\nu n_\nu - \sum_{\nu=1}^{d} \beta_\nu E_\nu, \tag{4.86}$$

where we have neglected contributions that arise from the change of the system's internal state: these contributions vanish anyway when identical initial and final states are considered, and for finite-sized systems they are negligibly small for large times, where the exchanged matter and energy contributions are dominating. For rate equations obeying local detailed balance (4.43), it can be shown quite generally that the characteristic polynomial of the rate matrix

$$\mathscr{D}(\chi,\xi) = \left| \mathscr{L}(\chi,\xi) - \lambda \mathbf{1} \right| \tag{4.87}$$

obeys the same symmetry, which then transfers to all eigenvalues of the Liouvillian and thus to the cumulant-generating function, too. Essentially, the proof [13] relies on analysis of the characteristic polynomial with Schnakenberg graph theory [14], but similar results may also be obtained with other methods [15, 16]. In particular, one obtains for d terminals with temperatures β_i and chemical potentials μ_i the shift relation (4.84) with

$$\Delta_\chi = (\beta_1\mu_1,\dots,\beta_d\mu_d), \qquad \Delta_\xi = (\beta_1,\dots,\beta_d). \tag{4.88}$$

In the long-term limit, the transfer of matter and energy to the terminals can be linked to the entropy flow in Definition 4.3, which at steady state is balanced by the entropy production in Definition 4.4. Therefore, the resulting fluctuation theorem describes the long-term statistics of entropy production:

$$\frac{P_{+\Delta n,+\Delta E}}{P_{-\Delta n,-\Delta E}} = \exp\left\{ \sum_\nu (\beta_\nu \mu_\nu n_\nu - \beta_\nu E_\nu) \right\} \quad \Leftrightarrow \quad \frac{P_{+\Delta_i S}}{P_{-\Delta_i S}} = e^{\Delta_i S} \tag{4.89}$$

and is a manifestation of the second law far from thermal equilibrium: trajectories with a negative entropy production are not completely forbidden but rather strongly suppressed, and it is straightforward to see that, on average, entropy production will always be positive. We show this by averaging over all trajectories:

$$\langle \Delta_i S \rangle = \sum_{\Delta_i S} \Delta_i S P_{\Delta_i S} = \sum_{\Delta_i S > 0} \Delta_i S (P_{+\Delta_i S} - P_{-\Delta_i S})$$

$$= \sum_{\Delta_i S > 0} \Delta_i S P_{-\Delta_i S} \underbrace{\left(e^{\Delta_i S} - 1 \right)}_{\geq 0} \geq 0. \tag{4.90}$$

Symmetries as in Eq. (4.87) hold in the rate equation (weak coupling) limit and imply of course that on average the second law is respected. The fluctuation relations have been verified, e.g., in an electronic setup [17, 18]. It turned out that slight modifications were visible, which can be explained by the interaction between system and detector. This interaction leads to further flows of information (physically connected to energy and matter flows) that modify the experimental signature.

It should be noted that when conservation laws exist, e.g. when the total particle current and/or the total energy current is conserved, the fluctuation theorem further simplifies. For example, for the SET we have conservation of the total particle number $n_L + n_R + n_d = \text{const.}$, where $n_d \in \{0, 1\}$ denotes the number of electrons on the dot. In the long-time limit, many particles will have been exchanged with the central dot of the SET and its terminals, and we will have in an approximate sense the conservation law $n_L = -n_R$. Furthermore, transferred energy and particles are tightly coupled in the master equation description, such that $\Delta E_\alpha = \Delta N_\alpha \varepsilon$ with dot level ε. Therefore, one can quantify the long-term entropy production simply by counting the number of particles transferring the SET, e.g., from left to right. Denoting the corresponding distribution by $P_n(t)$, the fluctuation theorem for equal temperatures simply becomes

$$\lim_{t \to \infty} \frac{P_{+n}}{P_{-n}} = e^{n\beta(\mu_L - \mu_R)}. \tag{4.91}$$

Exercise 4.9 (Fluctuation theorem) Find the fluctuation theorem, i.e., a symmetry in the cumulant-generating function, for the SET

$$\mathscr{L}(\chi) = \begin{pmatrix} -\Gamma_L f_L - \Gamma_R f_R & +\Gamma_L(1 - f_L) + \Gamma_R(1 - f_R)e^{+i\chi} \\ +\Gamma_L f_L + \Gamma_R f_R e^{-i\chi} & -\Gamma_L(1 - f_L) - \Gamma_R(1 - f_R) \end{pmatrix}.$$

References

1. W.H. Press, S.A. Teukolsky, W.T. Vetterling, B.P. Flannery, *Numerical Recipes in C*, 2nd edn. (Cambridge University Press, Cambridge, 1994)
2. N. Gershenfeld, *The Nature of Mathematical Modeling* (Cambridge University Press, Cambridge, 2000)
3. H.-P. Breuer, F. Petruccione, *The Theory of Open Quantum Systems* (Oxford University Press, Oxford, 2002)
4. H.-P. Breuer, Genuine quantum trajectories for non-Markovian processes. Phys. Rev. A **70**, 012106 (2004)

5. H.-P. Breuer, J. Piilo, Stochastic jump processes for non-Markovian quantum dynamics. Europhys. Lett. **85**, 50004 (2009)
6. M. Esposito, K. Lindenberg, C.V. den Broeck, Universality of efficiency at maximum power. Phys. Rev. Lett. **102**, 130602 (2009)
7. H.B. Callen, *Thermodynamics and an Introduction to Thermostatistics* (Wiley, New York, 1985)
8. T. Krause, G. Schaller, T. Brandes, Incomplete current fluctuation theorems for a four-terminal model. Phys. Rev. B **84**, 195113 (2011)
9. L. Simine, D. Segal, Vibrational cooling, heating, and instability in molecular conducting junctions: full counting statistics analysis. Phys. Chem. Chem. Phys. **14**, 13820 (2012)
10. A.N. Jordan, E.V. Sukhorukov, Transport statistics of bistable systems. Phys. Rev. Lett. **93**, 260604 (2004)
11. G. Schaller, G. Kießlich, T. Brandes, Counting statistics in multistable systems. Phys. Rev. B **81**, 205305 (2010)
12. U. Seifert, Entropy production along a stochastic trajectory and an integral fluctuation theorem. Phys. Rev. Lett. **95**, 040602 (2005)
13. D. Andrieux, P. Gaspard, Fluctuation theorem for currents and Schnakenberg network theory. J. Stat. Phys. **127**, 107 (2007)
14. J. Schnakenberg, Network theory of microscopic and macroscopic behavior of master equation systems. Rev. Mod. Phys. **48**, 571 (1976)
15. M. Esposito, U. Harbola, S. Mukamel, Nonequilibrium fluctuations, fluctuation theorems, and counting statistics in quantum systems. Rev. Mod. Phys. **81**, 1665 (2009)
16. M. Campisi, P. Hänggi, P. Talkner, Colloquium: quantum fluctuation relations: foundations and applications. Rev. Mod. Phys. **83**, 771 (2011)
17. Y. Utsumi, D.S. Golubev, M. Marthaler, K. Saito, T. Fujisawa, G. Schön, Bidirectional single-electron counting and the fluctuation theorem. Phys. Rev. B **81**, 125331 (2010)
18. S. Nakamura, Y. Yamauchi, M. Hashisaka, K. Chida, K. Kobayashi, T. Ono, R. Leturcq, K. Ensslin, Y. Saito, Y. Utsumi, A.C. Gossard, Nonequilibrium fluctuation relations in a quantum coherent conductor. Phys. Rev. Lett. **104**, 080602 (2010)

Chapter 5
Composite Non-equilibrium Environments

Abstract This chapter discusses models that assume a stationary non-equilibrium steady state without any external interventions. Mostly, these may be implemented by electronic setups, e.g., with transport through quantum dots or molecules, but the general machinery is also applicable to quantum optical setups. We first investigate the single electron transistor (SET) and afterwards the double quantum dot (DQD) with a focus on the thermodynamic interpretation. To mimic the interaction of such systems with a charge detector, we afterwards consider interacting transport channels: two coupled SETs and an SET coupled to a low-transparency quantum point contact (QPC). We discuss the decoherence of a charge qubit induced by a QPC. As an example for a setup where the environment itself is in a non-equilibrium steady state that cannot be expressed as a simple tensor product of different equilibrium states, we discuss an SET (which is solved exactly) weakly coupled to a DQD. We conclude by discussing models involving bosons and fermions simultaneously: this includes a model where phonon-assisted tunneling may be exploited to implement a thermoelectric generator. Finally, we present a model where—despite the strong coupling between the electronic system and the bosonic reservoir—a description within a simple rate equation is still possible. Despite its low dimensionality, the model allows for a rich dynamics and inspiring thermodynamic interpretation.

The simplest way to construct a non-equilibrium reservoir using existing knowledge is to envisage reservoirs that are composed of several different equilibrium components, e.g., held at different temperatures. Each component of the reservoir is then a separate equilibrium environment, and when these components are only indirectly coupled via a small quantum system, one may expect the resulting non-equilibrium state—which actually is composed of many equilibrium states—to persist. When the quantum system is coupled to all the components constructing its environment, it altogether experiences an environment that may be extremely far from equilibrium. Alternatively, we may treat environments that are in a non-equilibrium stationary state. Such non-equilibrium stationary states may be obtained as exact solutions of noninteracting models. Here, a product state might only be used as an initial state, and the final stationary state cannot simply be expressed as products of equilibrium

G. Schaller, *Open Quantum Systems Far from Equilibrium*, Lecture Notes in Physics 881,   87
DOI 10.1007/978-3-319-03877-3_5,
© Springer International Publishing Switzerland 2014

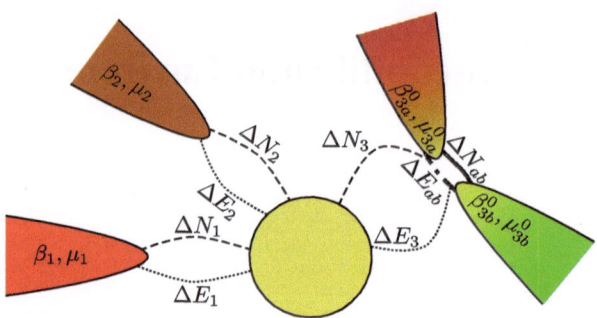

Fig. 5.1 Sketch of a system (*yellow circle*) that is coupled via weak energy (*dotted lines*) and weak particle (*dashed lines*) exchanges to a non-equilibrium environment. In the example shown, the first two components of the reservoir are thermal equilibrium states—fully characterized by temperatures and chemical potentials. In contrast, the coupling between reservoirs 3a and 3b admits strong energy (*dash–dotted*) and particle (*solid*) transfers. When these couplings are of similar order as the local Hamiltonians of the two leads, the third reservoir will assume a non-equilibrium stationary state, which can be found by solving for the isolated evolution of the third reservoir

states. In fact, it cannot be expected to be a thermal state at all. Figure 5.1 illustrates the difference between these two setups.

In the following, we will discuss several models of the above type.

5.1 Single Electron Transistor (SET)

A transistor conventionally has three terminals: a gate terminal can be used to tune the current through the other two terminals. Keeping the control gate implicit in the description, we describe the single electron transistor (SET) as a single quantum dot that is tunnel-coupled to two leads, i.e., with the Hamiltonian in Eq. (3.26); see also Fig. 5.2. Though the model is exactly solvable [1] (compare Sect. 3.2), the corresponding master equation that is valid for weak electronic tunneling amplitudes enables for interesting thermodynamic interpretations, and as one of the simplest possible non-equilibrium setups it will also be used as an introductory example here.

5.1.1 Model

To obtain the master equation, we can map the interaction Hamiltonian to a tensor product representation as described in the beginning of Sect. 2.1, such that, e.g., the part describing jumps to the left lead reads as

$$H_I^{(L)} = \tilde{d} \otimes \sum_k t_{kL} \tilde{c}_{kL}^\dagger + \tilde{d}^\dagger \otimes \sum_k t_{kL}^* \tilde{c}_{kL} = A_1 \otimes B_1 + A_2 \otimes B_2. \qquad (5.1)$$

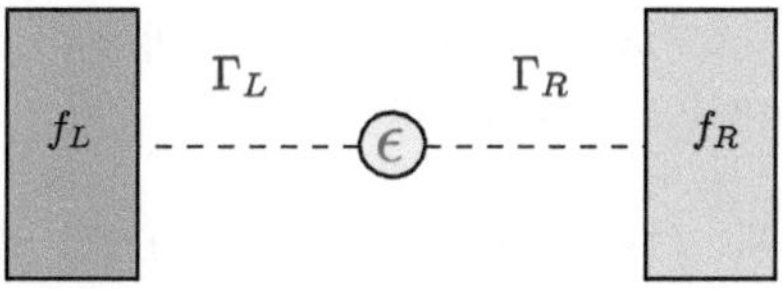

Fig. 5.2 Sketch of a single electron transistor (SET): a quantum dot with on-site energy ε that is weakly (*dashed lines*) tunnel-coupled with rates $\Gamma_{L/R}$ to two leads. These are held at thermal equilibrium and can thus be described by Fermi functions $f_{L/R}$. A third gate (not shown) can be used to tune the dot level ε and thereby the current through the SET

This representation yields for the bath correlation functions

$$
\begin{aligned}
C_{12}(t) &= \langle B_1(t) B_2 \rangle = \sum_{kk'} t_{kL} t_{k'L}^* e^{+i\varepsilon_{kL}t} \langle c_{kL}^\dagger c_{k'L} \rangle \\
&= \sum_k |t_{kL}|^2 e^{+i\varepsilon_{kL}t} f_L(\varepsilon_{kL}), \\
C_{21}(t) &= \langle B_2(t) B_1 \rangle = \sum_{kk'} t_{kL}^* t_{k'L} e^{-i\varepsilon_{kL}t} \langle c_{kL} c_{k'L}^\dagger \rangle \\
&= \sum_k |t_{kL}|^2 e^{-i\varepsilon_{kL}t} \left[1 - f_L(\varepsilon_{kL}) \right].
\end{aligned}
\tag{5.2}
$$

A reservoir has a continuum of frequencies and is thus infinitely large, which is a necessary condition for a stationary state to occur (finite-size quantum systems always evolve periodically). Formally, we take this into account by introducing the tunneling rates

$$
\Gamma_\alpha(\omega) = 2\pi \sum_k |t_{k\alpha}|^2 \delta(\omega - \varepsilon_{k\alpha}),
\tag{5.3}
$$

which allows us to write the summations in the correlation functions as integrals:

$$
\begin{aligned}
C_{12}(t) &= \frac{1}{2\pi} \int e^{+i\omega t} \Gamma_L(\omega) f_L(\omega) \, d\omega, \\
C_{21}(t) &= \frac{1}{2\pi} \int e^{-i\omega t} \Gamma_L(\omega) \left[1 - f_L(\omega) \right] d\omega.
\end{aligned}
\tag{5.4}
$$

These integral representations directly allow us to identify the Fourier transforms of the bath correlation functions without further calculations:

$$
\gamma_{12}(\omega) = \Gamma_L(-\omega) f_L(-\omega), \qquad \gamma_{21}(\omega) = \Gamma_L(+\omega) \left[1 - f_L(+\omega) \right].
\tag{5.5}
$$

Exactly the same treatment would follow for the coupling of the jumps associated to the right lead, such that we may transfer these results by replacing $L \to R$. A straightforward application of the quantum optical master equation in Definition 2.3 now yields in the energy eigenbasis $H_S|0\rangle = 0$ and $H_S|1\rangle = \varepsilon|1\rangle$ with $\langle 0|\tilde{d}|1\rangle = 1$ a rate equation for the populations,

$$\frac{d}{dt}\begin{pmatrix} P_0 \\ P_1 \end{pmatrix} = \sum_{\alpha \in \{L,R\}} \begin{pmatrix} -\Gamma_\alpha(\varepsilon)f_\alpha(\varepsilon) & +\Gamma_\alpha(\varepsilon)(1-f_\alpha(\varepsilon)) \\ +\Gamma_\alpha(\varepsilon)f_\alpha(\varepsilon) & -\Gamma_\alpha(\varepsilon)(1-f_\alpha(\varepsilon)) \end{pmatrix} \begin{pmatrix} P_0 \\ P_1 \end{pmatrix}, \qquad (5.6)$$

where $P_i = \rho_{ii}$ denote the probabilities of finding the dot empty or filled, respectively. Obviously, local detailed balance is fulfilled, since

$$\frac{\mathscr{L}_{EF}^{(\alpha)}}{\mathscr{L}_{FE}^{(\alpha)}} = \frac{1-f_\alpha(\varepsilon)}{f_\alpha(\varepsilon)} = e^{\beta_\alpha(\varepsilon-\mu_\alpha)}, \qquad (5.7)$$

which—if the system was only coupled to a single reservoir—would lead to equilibration of both system temperature and chemical potential with that of the reservoir $\bar{\rho} \to e^{-\beta(H_S-\mu N_S)}/Z_S$. The general stationary state is consistent with the occupation of the exact solution in the weak coupling limit, Eq. (3.42). It should be noted that it can be consistently expressed as a thermal state, i.e., by inverse temperature $\bar{\beta} > 0$ and a chemical potential $\bar{\mu}$, even if the terminals of the SET are far from equilibrium. This however results from the fact that the system—as any two-level system—only admits a single transition frequency.

It would of course be possible to formally calculate the decay of coherences, too. However, for the present system such coherences would correspond to superpositions of differently charged states, which are forbidden when the state of the full system is given by a tensor product of system and bath density matrices. For this model, such coherences can therefore not be created in the weak coupling limit, which already highlights one shortcoming of the master equation. The rates in the rate matrix (5.6) have a very simple interpretation: for example, the rate to tunnel into the dot from lead α is given by the bare tunneling rate $\Gamma_\alpha(\varepsilon)$ multiplied by the probability $f_\alpha(\varepsilon)$ to have an electron in the junction α at the required energy ε. The bare tunneling rate Γ_α results from the potential landscape between lead α and the system but does not depend on their occupations. The rate of the inverse process, i.e., tunneling from the system into lead α, is therefore given by a product of the bare tunneling rate and the probability $1 - f_\alpha(\varepsilon)$ to have a free space at the junction at energy ε, where the Pauli exclusion principle in the leads becomes manifest. When now the chemical potentials are different, it becomes obvious that in order to support a current through the dot, it is necessary that one lead has to provide electrons at the dot energy whereas the other lead must provide free space; see Fig. 5.3. In other words, the dot transition frequency must be situated within the transport window— which for sufficiently low lead temperatures means that it should be between μ_L and μ_R—to support a current. Since the electronic on-site energy ε, and thereby the current, may be tuned with a third gate, the setup is called a single electron transistor.

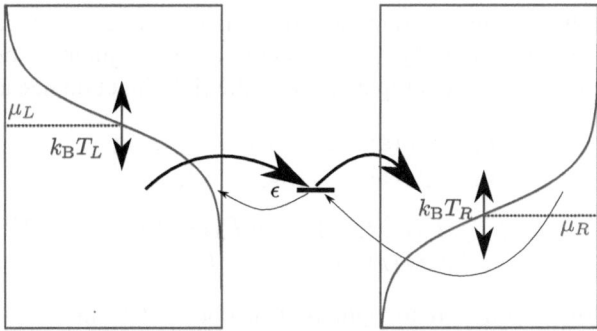

Fig. 5.3 Level sketch of a single quantum dot that is tunnel-coupled to two fermionic junctions, where the electronic occupation of energy levels is well approximated by a Fermi distribution. Within the master equation picture, transport from left to right is only enabled when—at the dot transition frequency—the left lead is occupied and the right one is empty

5.1.2 Thermodynamic Interpretation

Since we can directly identify the electronic jumps in the rate equation, it is straight-forward to track the energy and particle transfers occurring with each jump. In particular, the model exhibits what is called the tight coupling property: every electronic jump transports the same amount of energy ε, since the system only allows for a single transition frequency. Written in terms of energy and matter currents, this yields $I_E^{(\nu)} = \varepsilon I_M^{(\nu)}$. Furthermore, we can use the fact that energy and matter currents are conserved $I_E^{(L)} + I_E^{(R)} = 0$ and $I_M^{(L)} + I_M^{(R)} = 0$. At steady state, we can express the entropy production with the entropy flow

$$
\begin{aligned}
\dot{S}_{\mathrm{i}} &= -\dot{S}_{\mathrm{e}} = -\beta_L \dot{Q}^{(L)} - \beta_R \dot{Q}^{(R)} \\
&= -\beta_L \big(\varepsilon I_M^{(L)} - \mu_L I_M^{(L)}\big) - \beta_R \big(-\varepsilon I_M^{(L)} + \mu_R I_M^{(L)}\big) \\
&= \big[(\beta_R - \beta_L)\varepsilon + (\beta_L \mu_L - \beta_R \mu_R)\big] I_M^{(L)},
\end{aligned}
\tag{5.8}
$$

which is thus just proportional to the matter current through the SET (counted positive when traversing the system from left to right). At equal temperatures $\beta_L = \beta_R = \beta$, we thus see that the second law $\dot{S}_{\mathrm{i}} \geq 0$ requires the current to flow from the junction with higher chemical potential to the junction with lower chemical potential. At equal chemical potentials $\mu_L = \mu_R = \mu$ but different temperatures, the average heat current $\dot{Q}^{(L)} = (\varepsilon - \mu) I_M^{(L)}$ will flow from the hot junction to the cold junction. Furthermore, we note that a temperature gradient can in principle also drive the average current against a potential gradient, such that one might also see the SET as a thermoelectric device that converts a thermal gradient into useful power [2].

Since for long times, the entropy production can be quantified by just tracking the matter current through the SET, the matter current $I_M^{(L)}$ may be quantified by counting electrons jumping in from the left positively and electrons jumping out to the left negatively. Alternatively, we may count electrons jumping out to the right terminal positively and those jumping in from the right junction negatively:

$$\mathcal{L}(\chi) = \begin{pmatrix} -\Gamma_L(\varepsilon)f_L(\varepsilon) & +\Gamma_L(\varepsilon)(1 - f_L(\varepsilon)) \\ +\Gamma_L(\varepsilon)f_L(\varepsilon) & -\Gamma_L(\varepsilon)(1 - f_L(\varepsilon)) \end{pmatrix}$$

$$+ \begin{pmatrix} -\Gamma_R(\varepsilon)f_R(\varepsilon) & +\Gamma_R(\varepsilon)(1 - f_R(\varepsilon))e^{+i\chi} \\ +\Gamma_R(\varepsilon)f_R(\varepsilon)e^{-i\chi} & -\Gamma_R(\varepsilon)(1 - f_R(\varepsilon)) \end{pmatrix}. \tag{5.9}$$

The matter current from left to right is then obtained from the stationary state $\mathcal{L}(0)\bar{\rho} = 0$ via

$$I_M^{(L)} = -i\,\mathrm{Tr}\{\mathcal{L}'(0)\bar{\rho}\} = \frac{\Gamma_L(\varepsilon)\Gamma_R(\varepsilon)}{\Gamma_L(\varepsilon) + \Gamma_R(\varepsilon)}\big[f_L(\varepsilon) - f_R(\varepsilon)\big]. \tag{5.10}$$

This expression perfectly agrees with the exact solution of the model in Eq. (3.50) obtained for the wide-band limit $\Gamma_\alpha(\varepsilon) = \Gamma_\alpha$. Just for brevity omitting the dot-level dependence in Fermi functions and tunneling rates, the characteristic polynomial of the rate matrix becomes

$$\mathcal{D}(\chi) = \big|\mathcal{L}(\chi) - \lambda \mathbf{1}\big|$$
$$= \Gamma_L \Gamma_R (f_L + f_R - 2f_L f_R) + \lambda(\Gamma_L + \Gamma_R + \lambda)$$
$$- e^{+i\chi}\Gamma_L\Gamma_R f_L(1 - f_R) - e^{-i\chi}\Gamma_L\Gamma_R(1 - f_L)f_R, \tag{5.11}$$

and we can immediately verify the symmetry $\mathcal{D}(-\chi) = \mathcal{D}(+\chi + i\mathcal{A})$ with the affinity

$$\mathcal{A} = \ln\left(\frac{(1 - f_L)f_R}{f_L(1 - f_R)}\right) = (\beta_R - \beta_L)\varepsilon + (\beta_L\mu_L - \beta_R\mu_R)$$
$$\stackrel{\beta_L = \beta_R = \beta}{\longrightarrow} \beta(\mu_L - \mu_R), \tag{5.12}$$

which transfers to the eigenvalues of the rate matrix and thus to the long-term cumulant-generating function. Denoting the probability of transferring n particles from left to right through the SET after time t by $P_n(t)$, this symmetry therefore implies a fluctuation theorem:

$$\lim_{t\to\infty} \frac{P_{+n}(t)}{P_{-n}(t)} = e^{n\mathcal{A}} = e^{[(\beta_R - \beta_L)\varepsilon + (\beta_L\mu_L - \beta_R\mu_R)]n}. \tag{5.13}$$

We note that the affinity in Eq. (5.12) is also occurring in the average entropy production (5.8). This is not unexpected, as the fluctuation theorem (5.13) relates the probabilities of forward trajectories $P_{+n}(t)$ with a corresponding entropy production $\Delta_i S \approx [(\beta_R - \beta_L)\varepsilon + (\beta_L\mu_L - \beta_R\mu_R)]n$ with the probability for the corresponding backward trajectory $P_{-n}(t)$. Thus, whereas Eq. (5.8) expresses the second law at the average level, Eq. (5.13) even quantifies the second law at the trajectory level.

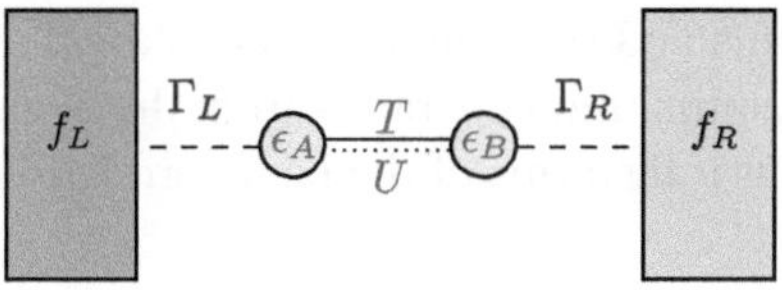

Fig. 5.4 A double quantum dot (system) with on-site energies $\varepsilon_{A/B}$ and internal tunneling amplitude T (*solid line*) and Coulomb interaction U (*dotted line*) may host at most two electrons. It is weakly (*dashed lines*) tunnel-coupled to two fermionic contacts via the rates $\Gamma_{L/R}$ at different thermal equilibria described by the Fermi distributions $f_{L/R}(\omega)$. In the rate equation, these are just evaluated at the transition energies of the system and can be characterized by the temperatures and chemical potentials of the leads

5.2 Serial Double Quantum Dot

To advance to a system with more than just a single transition frequency, we consider a double quantum dot (DQD) with internal tunnel amplitude T and Coulomb interaction U that is weakly coupled to two fermionic contacts via the rates Γ_L and Γ_R. See Fig. 5.4.

5.2.1 Model

The corresponding Hamiltonian reads

$$H_S = \varepsilon_A d_A^\dagger d_A + \varepsilon_B d_B^\dagger d_B + T\left(d_A d_B^\dagger + d_B d_A^\dagger\right) + U d_A^\dagger d_A d_B^\dagger d_B,$$

$$H_B = \sum_k \varepsilon_{kL} c_{kL}^\dagger c_{kL} + \sum_k \varepsilon_{kR} c_{kR}^\dagger c_{kR}, \qquad (5.14)$$

$$H_I = \sum_k \left(t_{kL} d_A c_{kL}^\dagger + t_{kL}^* c_{kL} d_A^\dagger\right) + \sum_k \left(t_{kR} d_B c_{kR}^\dagger + t_{kR}^* c_{kR} d_B^\dagger\right).$$

Note that initially we do not have a tensor product decomposition in the interaction Hamiltonian, as the original coupling operators anti-commute $\{d, c_{kR}\} = 0$. We may however use the Jordan–Wigner transform discussed in Sect. 2.1 to map to fermions that are separately defined on system and bath. In the new operator basis, the Hamiltonian appears as

$$H_S = \left[\varepsilon_A \tilde{d}_A^\dagger \tilde{d}_A + \varepsilon_B \tilde{d}_B^\dagger \tilde{d}_B + T\left(\tilde{d}_A \tilde{d}_B^\dagger + \tilde{d}_B \tilde{d}_A^\dagger\right) + U \tilde{d}_A^\dagger \tilde{d}_A \tilde{d}_B^\dagger \tilde{d}_B\right] \otimes \mathbf{1},$$

$$H_B = \mathbf{1} \otimes \left[\sum_k \varepsilon_{kL} \tilde{c}_{kL}^\dagger \tilde{c}_{kL} + \sum_k \varepsilon_{kR} \tilde{c}_{kR}^\dagger \tilde{c}_{kR}\right], \qquad (5.15)$$

$$H_I = \tilde{d}_A \otimes \sum_k t_{kL} \tilde{c}_{kL}^\dagger + \tilde{d}_A^\dagger \otimes \sum_k t_{kL}^* \tilde{c}_{kL} + \tilde{d}_B \otimes \sum_k t_{kR} \tilde{c}_{kR}^\dagger + \tilde{d}_B^\dagger \otimes \sum_k t_{kR}^* \tilde{c}_{kR},$$

which is the same (for this and some more special cases) as if we had ignored the fermionic nature of the annihilation operators from the beginning. We now proceed by calculating the Fourier transforms of the bath correlation functions as we did in the previous chapter,

$$
\begin{aligned}
\gamma_{12}(\omega) &= \Gamma_L(-\omega) f_L(-\omega), & \gamma_{21}(\omega) &= \Gamma_L(+\omega)\big[1 - f_L(+\omega)\big], \\
\gamma_{34}(\omega) &= \Gamma_R(-\omega) f_R(-\omega), & \gamma_{43}(\omega) &= \Gamma_R(+\omega)\big[1 - f_R(+\omega)\big]
\end{aligned}
\tag{5.16}
$$

with the continuum tunneling rates $\Gamma_\alpha(\omega) = 2\pi \sum_k |t_{k\alpha}|^2 \delta(\omega - \varepsilon_{k\alpha})$ and Fermi functions $f_\alpha(\varepsilon_{k\alpha}) = \langle c_{k\alpha}^\dagger c_{k\alpha} \rangle$.

Exercise 5.1 (DQD bath correlation functions) Calculate the Fourier transforms (5.16) of the bath correlation functions for the DQD.

Next, we diagonalize the system Hamiltonian (in the Fock space basis)

$$
\begin{aligned}
E_0 &= 0, & |v_0\rangle &= |00\rangle, \\
E_- &= \varepsilon - \sqrt{\Delta^2 + T^2}, & |v_-\rangle &\propto \big[-(\Delta + \sqrt{\Delta^2 + T^2})|10\rangle + T|01\rangle\big], \\
E_+ &= \varepsilon + \sqrt{\Delta^2 + T^2}, & |v_+\rangle &\propto \big[-(\Delta - \sqrt{\Delta^2 + T^2})|10\rangle + T|01\rangle\big], \\
E_2 &= 2\varepsilon + U, & |v_2\rangle &= |11\rangle,
\end{aligned}
\tag{5.17}
$$

where $\Delta = (\varepsilon_B - \varepsilon_A)/2$ and $\varepsilon = (\varepsilon_A + \varepsilon_B)/2$ and $|01\rangle = \tilde{d}_B^\dagger |00\rangle$, $|10\rangle = \tilde{d}_A^\dagger |00\rangle$, and $|11\rangle = \tilde{d}_B^\dagger \tilde{d}_A^\dagger |00\rangle$. To obtain the quantum optical master equation we may use Definition 2.3. Specifically, when we have no degeneracies in the system Hamiltonian ($\Delta^2 + T^2 > 0$), the master equation in the energy eigenbasis (where $a, b \in \{0, -, +, 2\}$) becomes a rate equation $\dot{\rho}_{aa} = +\sum_b \gamma_{ab,ab} \rho_{bb} - [\sum_b \gamma_{ba,ba}]\rho_{aa}$ with the transition rate from state b to state a,

$$
\gamma_{ab,ab} = \sum_{\alpha\beta} \gamma_{\alpha\beta}(E_b - E_a)\langle a|A_\beta|b\rangle \langle a|A_\alpha^\dagger|b\rangle^*.
\tag{5.18}
$$

As a consequence of the weak coupling limit, there are no correlations between left and right terminals, so we may calculate the Liouvillians for the interaction with the left and right contact separately:

$$
\gamma_{ab,ab} = \gamma_{ab,ab}^L + \gamma_{ab,ab}^R.
\tag{5.19}
$$

Such simple additive decompositions are generally expected in the weak coupling (or sequential tunneling) limit but may not hold beyond these limits [3]. The corresponding rate equation $(\dot{P}_0, \dot{P}_-, \dot{P}_+, \dot{P}_2)^T = \mathscr{L}(P_0, P_-, P_+, P_2)^T$ with the rate

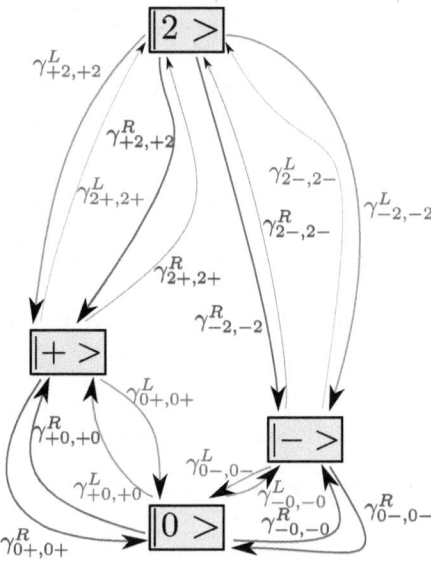

Fig. 5.5 Configuration space of a serial double quantum dot coupled to two leads. Due to the hybridization of the two levels, electrons may jump directly from the left contact to right-localized modes and vice versa, such that in principle all transitions are driven by both contacts. However, the relative strength of the couplings is different, such that the two Liouvillians have a different structure. In the Coulomb blockade limit, transitions to the doubly occupied state are strongly suppressed (*thin lines*), such that the dimension of the rate equation can be reduced

matrix

$$
\mathscr{L} = \begin{pmatrix}
-\gamma_{-0,-0} - \gamma_{+0,+0} & \gamma_{0-,0-} & \gamma_{0+,0+} & 0 \\
\gamma_{-0,-0} & -\gamma_{0-,0-} - \gamma_{2-,2-} & 0 & \gamma_{-2,-2} \\
\gamma_{+0,+0} & 0 & -\gamma_{0+,0+} - \gamma_{2+,2+} & \gamma_{+2,+2} \\
0 & \gamma_{2-,2-} & \gamma_{2+,2+} & -\gamma_{-2,-2} - \gamma_{+2,+2}
\end{pmatrix}
$$

(5.20)

only allows transitions changing the system charge by one. Such rate matrices can also be visualized with a network; see Fig. 5.5. Since we have $\tilde{d}_A = A_2^\dagger = A_1$ and $\tilde{d}_B = A_4^\dagger = A_2$, we obtain for the damping coefficients

$$
\gamma_{ab,ab}^L = \gamma_{12}(E_b - E_a)|\langle a|A_2|b\rangle|^2 + \gamma_{21}(E_b - E_a)|\langle a|A_1|b\rangle|^2,
$$
$$
\gamma_{ab,ab}^R = \gamma_{34}(E_b - E_a)|\langle a|A_4|b\rangle|^2 + \gamma_{43}(E_b - E_a)|\langle a|A_3|b\rangle|^2.
$$

(5.21)

In the energy eigenbasis, the rates therefore become (just for brevity, in the wide-band limit $\Gamma_\alpha(\omega) = \Gamma_\alpha$):

$$
\gamma_{0-,0-}^L = \Gamma_L \gamma_+ \left[1 - f_L\left(\varepsilon - \sqrt{\Delta^2 + T^2}\right)\right],
$$
$$
\gamma_{0-,0-}^R = \Gamma_R \gamma_- \left[1 - f_R\left(\varepsilon - \sqrt{\Delta^2 + T^2}\right)\right],
$$

$$\gamma^L_{0+,0+} = \Gamma_L \gamma_- \left[1 - f_L \left(\varepsilon + \sqrt{\Delta^2 + T^2} \right) \right],$$

$$\gamma^R_{0+,0+} = \Gamma_R \gamma_+ \left[1 - f_R \left(\varepsilon + \sqrt{\Delta^2 + T^2} \right) \right],$$

$$\gamma^L_{-2,-2} = \Gamma_L \gamma_- \left[1 - f_L \left(\varepsilon + U + \sqrt{\Delta^2 + T^2} \right) \right],$$

$$\gamma^R_{-2,-2} = \Gamma_R \gamma_+ \left[1 - f_R \left(\varepsilon + U + \sqrt{\Delta^2 + T^2} \right) \right],$$

$$\gamma^L_{+2,+2} = \Gamma_L \gamma_+ \left[1 - f_L \left(\varepsilon + U - \sqrt{\Delta^2 + T^2} \right) \right],$$

$$\gamma^R_{+2,+2} = \Gamma_R \gamma_- \left[1 - f_R \left(\varepsilon + U - \sqrt{\Delta^2 + T^2} \right) \right],$$

$$\gamma^L_{-0,-0} = \Gamma_L \gamma_+ f_L \left(\varepsilon - \sqrt{\Delta^2 + T^2} \right),$$

$$\gamma^R_{-0,-0} = \Gamma_R \gamma_- f_R \left(\varepsilon - \sqrt{\Delta^2 + T^2} \right),$$

$$\gamma^L_{+0,+0} = \Gamma_L \gamma_- f_L \left(\varepsilon + \sqrt{\Delta^2 + T^2} \right),$$

$$\gamma^R_{+0,+0} = \Gamma_R \gamma_+ f_R \left(\varepsilon + \sqrt{\Delta^2 + T^2} \right),$$

$$\gamma^L_{2-,2-} = \Gamma_L \gamma_- f_L \left(\varepsilon + U + \sqrt{\Delta^2 + T^2} \right),$$

$$\gamma^R_{2-,2-} = \Gamma_R \gamma_+ f_R \left(\varepsilon + U + \sqrt{\Delta^2 + T^2} \right),$$

$$\gamma^L_{2+,2+} = \Gamma_L \gamma_+ f_L \left(\varepsilon + U - \sqrt{\Delta^2 + T^2} \right),$$

$$\gamma^R_{2+,2+} = \Gamma_R \gamma_- f_R \left(\varepsilon + U - \sqrt{\Delta^2 + T^2} \right),$$

$$(5.22)$$

with the dimensionless but positive coefficients

$$\gamma_\pm = \frac{1}{2} \left[1 \pm \frac{\Delta}{\sqrt{\Delta^2 + T^2}} \right] \tag{5.23}$$

arising from the matrix elements of the system coupling operators. We note that local detailed balance relations are obeyed:

$$\frac{\gamma^\alpha_{ab,ab}}{\gamma^\alpha_{ba,ba}} = \frac{1 - f_\alpha(E_b - E_a)}{f_\alpha(E_b - E_a)}. \tag{5.24}$$

As the simplest example of the resulting rate equation, we study the Coulomb blockade limit: for large Coulomb interactions U, one will obtain $f_{L/R}(\varepsilon + U \pm \sqrt{\Delta^2 + T^2}) \to 0$. If this is combined with a high-bias assumption $f_L(\varepsilon \pm \sqrt{\Delta^2 + T^2}) \to 1$ and $f_R(\varepsilon \pm \sqrt{\Delta^2 + T^2}) \to 0$, the rates simplify considerably. Furthermore, we assume for simplicity $\Delta \to 0$ (such that $\gamma_\pm \to 1/2$). The resulting Liouvillian reads as

$$\mathscr{L} = \frac{1}{2} \begin{pmatrix} -2\Gamma_L & \Gamma_R & \Gamma_R & 0 \\ \Gamma_L & -\Gamma_R & 0 & \Gamma_L + \Gamma_R \\ \Gamma_L & 0 & -\Gamma_R & \Gamma_L + \Gamma_R \\ 0 & 0 & 0 & -2(\Gamma_L + \Gamma_R) \end{pmatrix}, \tag{5.25}$$

where it becomes visible that the doubly occupied state will simply decay and may therefore—since we are interested in the long-term dynamics—be eliminated completely:

$$\mathscr{L}_{\text{CBHB}} = \frac{1}{2} \begin{pmatrix} -2\Gamma_L & \Gamma_R & \Gamma_R \\ \Gamma_L & -\Gamma_R & 0 \\ \Gamma_L & 0 & -\Gamma_R \end{pmatrix}. \tag{5.26}$$

Another often-studied limit is the infinite bias limit. Here, one assumes that the chemical potential of the source junction is much larger than all system transition frequencies and that temperature is finite, such that, e.g., $f_L(\omega) \to 1$. Similarly, the chemical potential of the drain junction is assumed to be much smaller than all transition frequencies, which would, e.g., imply that $f_R(\omega) \to 0$. Assuming for simplicity again that $\Delta = 0$, the Liouvillian in the infinite bias limit becomes

$$\mathscr{L}_{\text{IB}} = \frac{1}{2} \begin{pmatrix} -2\Gamma_L & \Gamma_R & \Gamma_R & 0 \\ \Gamma_L & -\Gamma_L - \Gamma_R & 0 & \Gamma_R \\ \Gamma_L & 0 & -\Gamma_L - \Gamma_R & \Gamma_R \\ 0 & +\Gamma_L & +\Gamma_L & -2\Gamma_R \end{pmatrix}. \tag{5.27}$$

The following warning is applicable: in the discussed limits, the Liouvillian does not depend on T. Consequently, an unphysical current would be predicted through a disconnected structure when $T \to 0$. However, precisely in this limit, the two levels E_- and E_+ become energetically degenerate when $T \to 0$, such that a simple rate equation description is not applicable. The Liouvillians (5.26) and (5.27) can only be expected to be valid when $T \gg \Gamma_\alpha$.

Exercise 5.2 (Stationary current) Calculate the stationary currents corresponding to rate matrices Eq. (5.26) and Eq. (5.27).

Exercise 5.3 (Non-equilibrium stationary state) Show that the stationary state of Eq. (5.26) cannot be written as a grand-canonical equilibrium state by disproving the equations $\bar{\rho}_{--}/\bar{\rho}_{00} = e^{-\beta(E_- - E_0 - \mu)}$, $\bar{\rho}_{++}/\bar{\rho}_{00} = e^{-\beta(E_+ - E_0 - \mu)}$, and $\bar{\rho}_{++}/\bar{\rho}_{--} = e^{-\beta(E_+ - E_-)}$.

The matter and energy currents through the DQD can be used to probe the system transition frequencies; see Fig. 5.6.

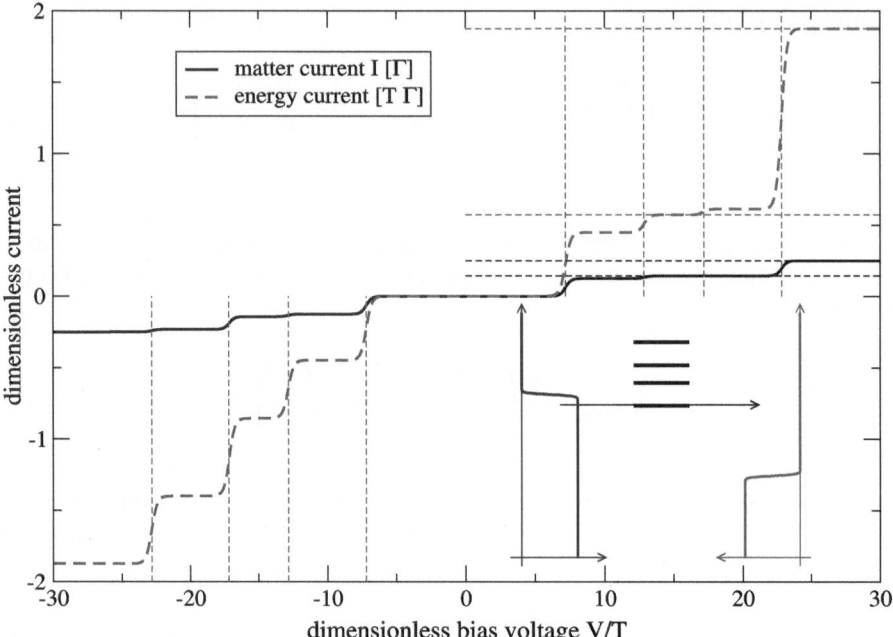

Fig. 5.6 Plot of matter (*solid black*, in units of $\Gamma_L = \Gamma_R = \Gamma$) and energy (*dashed red*, in units of ΓT) currents through the double quantum dot. At sufficiently low temperatures, the steps in the currents occur for positive bias voltage at $\mu_L = V/2 \in \{E_- - E_0, E_+ - E_0, E_2 - E_+, E_2 - E_-\}$ (compare *vertical dashed lines*). *Horizontal dashed lines* represent energy and matter currents in high-bias and Coulomb blockade limits (*lower lines*) and infinite bias limit (*upper lines*), respectively. The *inset* displays the configuration of these transition energies relative to left (*blue*) and right (*green*) Fermi functions taken at $V = 10T$. Then, only the lowest transition energy (*arrow*) is inside the transport window, such that transport is dominated by transitions between $|-\rangle$ and $|0\rangle$. Other parameters have been chosen as $\mu_L = -\mu_R = V/2$, $\Gamma_L = \Gamma_R = \Gamma$, $\varepsilon_A = 4T$, $\varepsilon_B = 6T$, $U = 5T$, and $\beta T = 10$

5.2.2 Thermodynamic Interpretation

Unlike the model in the previous section, this model no longer supports tight coupling, as the transferred energy may now depend on the particular jumps. Using energy and matter conservation, we can write the long-term entropy production as

$$\dot{S}_\text{i} = -\beta_L \dot{Q}^{(L)} - \beta_R \dot{Q}^{(R)} = (\beta_R - \beta_L)I_E^{(L)} + (\beta_L \mu_L - \beta_R \mu_R)I_M^{(L)}, \quad (5.28)$$

and we can interpret this expression in a similar manner as before.

To track energy and matter currents, we introduce two counting fields by performing in off-diagonal matrix elements of the rate matrix (5.20) the replacements

$$\gamma_{0-,0-}^{(L)} \rightarrow \gamma_{0-,0-}^{(L)} e^{-i\chi} e^{-i\xi(E_+ - E_0)},$$

$$\gamma_{0+,0+}^{(L)} \rightarrow \gamma_{0+,0+}^{(L)} e^{-i\chi} e^{-i\xi(E_+ - E_0)},$$

$$\gamma_{-0,-0}^{(L)} \rightarrow \gamma_{-0,-0}^{(L)} e^{+i\chi} e^{+i\xi(E_- - E_0)},$$

$$\gamma_{-2,-2}^{(L)} \rightarrow \gamma_{-2,-2}^{(L)} e^{-i\chi} e^{-i\xi(E_2 - E_-)},$$

$$\gamma_{+0,+0}^{(L)} \rightarrow \gamma_{+0,+0}^{(L)} e^{+i\chi} e^{+i\xi(E_+ - E_0)},$$

$$\gamma_{+2,+2}^{(L)} \rightarrow \gamma_{+2,+2}^{(L)} e^{-i\chi} e^{-i\xi(E_2 - E_+)},$$

$$\gamma_{2-,2-}^{(L)} \rightarrow \gamma_{2-,2-}^{(L)} e^{+i\chi} e^{+i\xi(E_2 - E_-)},$$

$$\gamma_{2+,2+}^{(L)} \rightarrow \gamma_{2+,2+}^{(L)} e^{+i\chi} e^{+i\xi(E_2 - E_+)}.$$

(5.29)

This yields a conditional rate equation with two counting fields $\mathscr{L}(\chi, \xi)$. By computing the characteristic polynomial $\mathscr{D}(\chi, \xi) = |\mathscr{L}(\chi, \xi) - \mathbf{1}|$ we can find the symmetry

$$\mathscr{D}(-\chi, -\xi) = \mathscr{D}(+\chi + i\mathscr{A}_M, +\xi + i\mathscr{A}_E) \qquad (5.30)$$

with the affinities for matter and energy flow

$$\mathscr{A}_M = \beta_L \mu_L - \beta_R \mu_R, \qquad \mathscr{A}_E = \beta_R - \beta_L. \qquad (5.31)$$

These imply a combined fluctuation theorem for energy and matter exchange. Denoting the probability to transfer n electrons together with the energy E from left to right after time t by $P_{n,E}(t)$, it reads

$$\lim_{t \to \infty} \frac{P_{+n, +E}(t)}{P_{-n, -E}(t)} = e^{(\beta_L \mu_L - \beta_R \mu_R)n + (\beta_R - \beta_L)E}. \qquad (5.32)$$

We note that if one is only able to count particles, similar fluctuation theorems would emerge if one would separately count the particles carrying a specific transition energy [4]. For example, counting all four transitions separately such that $n = n_1 + n_2 + n_3 + n_4$ and $E = (E_- - E_0)n_1 + (E_+ - E_0)n_2 + (E_2 - E_+)n_3 + (E_2 - E_-)n_4$ would yield the same fluctuation theorem.

5.3 Interacting Transport Channels: Two Coupled SETs

We consider two SETs A and B with the two circuits additionally obeying a capacitive interaction as depicted in Fig. 5.7.

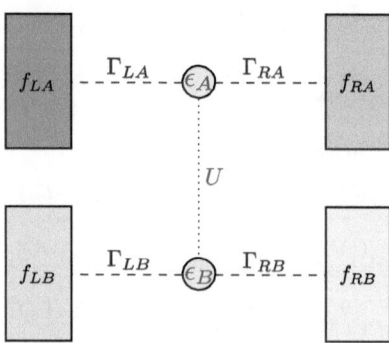

Fig. 5.7 Sketch of two SET circuits A and B that are capacitively coupled by the additional Coulomb interaction U (*dotted line*), which admits the exchange of energy between the two circuits. *Dashed lines* represent weak tunnel couplings with rate $\Gamma_{\alpha\beta}$ to the leads. These are assumed at separate thermal equilibrium and thus described by Fermi functions $f_{\alpha\beta}$. When, e.g., circuit B is much faster than circuit A, the current through B may rapidly adapt to the slow changes of the occupations of dot A. In this limit, channel B may be used as an electrometer for the counting statistics of channel A

5.3.1 Model

The corresponding Hamiltonian reads

$$
\begin{aligned}
H_S &= \varepsilon_A d_A^\dagger d_A + \varepsilon_B d_B^\dagger d_B + U d_A^\dagger d_A d_B^\dagger d_B, \\
H_B &= \sum_k \sum_{\alpha\in\{L,R\}} \sum_{\beta\in\{A,B\}} \varepsilon_{k\alpha\beta} c_{k\alpha\beta}^\dagger c_{k\alpha\beta}, \\
H_I &= \sum_k \sum_{\alpha\in\{L,R\}} \sum_{\beta\in\{A,B\}} \left(t_{k\alpha\beta} d_\beta c_{k\alpha\beta}^\dagger + t_{k\alpha\beta}^* c_{k\alpha\beta} d_\beta^\dagger \right),
\end{aligned}
\tag{5.33}
$$

and it becomes immediately obvious that the system energy eigenbasis coincides with the localized Fock state basis $|n_A, n_B\rangle$ with the occupations $n_\beta \in \{0, 1\}$. Due to the simplicity of the model, we may directly proceed to the rate equation in the system energy eigenbasis (which we deliberately order as $|00\rangle$, $|01\rangle$, $|10\rangle$, and $|11\rangle$). The rate matrix decomposes additively in all four terminals $\mathscr{L} = \mathscr{L}_{LA} + \mathscr{L}_{RA} + \mathscr{L}_{LB} + \mathscr{L}_{RB}$, where the separate contributions read

$$
\mathscr{L}_{\alpha A} = \begin{pmatrix}
-\Gamma_{\alpha A} f_{\alpha A} & 0 & +\Gamma_{\alpha A}(1-f_{\alpha A}) & 0 \\
0 & -\Gamma_{\alpha A}^U f_{\alpha A}^U & 0 & +\Gamma_{\alpha A}^U(1-f_{\alpha A}^U) \\
+\Gamma_{\alpha A} f_{\alpha A} & 0 & -\Gamma_{\alpha A}(1-f_{\alpha A}) & 0 \\
0 & +\Gamma_{\alpha A}^U f_{\alpha A}^U & 0 & -\Gamma_{\alpha A}^U(1-f_{\alpha A}^U)
\end{pmatrix},
$$

$$
\mathscr{L}_{\alpha B} = \begin{pmatrix}
-\Gamma_{\alpha B} f_{\alpha B} & +\Gamma_{\alpha B}(1-f_{\alpha B}) & 0 & 0 \\
+\Gamma_{\alpha B} f_{\alpha B} & -\Gamma_{\alpha B}(1-f_{\alpha B}) & 0 & 0 \\
0 & 0 & -\Gamma_{\alpha B}^U f_{\alpha B}^U & +\Gamma_{\alpha B}^U(1-f_{\alpha B}^U) \\
0 & 0 & +\Gamma_{\alpha B}^U f_{\alpha B}^U & -\Gamma_{\alpha B}^U(1-f_{\alpha B}^U)
\end{pmatrix}.
\tag{5.34}
$$

Since there are only four allowed transition frequencies in this model, we have introduced the abbreviations

$$\Gamma_{\alpha A} = \Gamma_{\alpha A}(\varepsilon_A), \qquad \Gamma_{\alpha A}^U = \Gamma_{\alpha A}(\varepsilon_A + U),$$
$$\Gamma_{\alpha B} = \Gamma_{\alpha B}(\varepsilon_B), \qquad \Gamma_{\alpha B}^U = \Gamma_{\alpha B}(\varepsilon_B + U),$$
$$f_{\alpha A} = f_{\alpha A}(\varepsilon_A), \qquad f_{\alpha A}^U = f_{\alpha A}(\varepsilon_A + U),$$
$$f_{\alpha B} = f_{\alpha B}(\varepsilon_B), \qquad f_{\alpha B}^U = f_{\alpha B}(\varepsilon_B + U),$$

$$(5.35)$$

linking to the tunneling rates $\Gamma_{\alpha\beta}(\omega)$ and Fermi functions $f_{\alpha\beta}(\omega)$ of each terminal. We note that local detailed balance is obeyed.

As before, the counting statistics of the model may yield system-specific features.

Exercise 5.4 (Transition rates) Derive the Fourier transforms of the reservoir correlation functions and confirm the rates in the Liouvillian (5.34).

In the limit when one channel is much slower than the other (e.g., $\Gamma_{\alpha A} \to 0$ and $\Gamma_{\alpha A}^U \to 0$), the model becomes bistable, as is in a suitable basis immediately evident from the block structure of the Liouvillian. In the extreme limit, the decoupled dot will just maintain its initial occupation, and the remaining channel just corresponds to an SET with a gate voltage that is tuned according to the (stationary) occupation of the decoupled dot. Consequently, the current through channel B at a fixed bias will depend on the initial occupation of channel A. Depending on the occupation of channel A, the current through B may assume the two values

$$I_1 = \frac{\Gamma_{LB}\Gamma_{RB}}{\Gamma_{LB} + \Gamma_{RB}}[f_{LB} - f_{RB}], \qquad I_2 = \frac{\Gamma_{LB}^U \Gamma_{RB}^U}{\Gamma_{LB}^U + \Gamma_{RB}^U}[f_{LB}^U - f_{RB}^U]. \quad (5.36)$$

When one has finite rates at both circuits but still $0 < \Gamma_{\alpha A} \ll \Gamma_{\alpha B}$, this still holds approximately: the fast channel B then switches (as dictated by the slow fluctuations of channel A) between the two values of the current, which is known as telegraph noise. Thus, the fast channel may serve as a detector reading out the instantaneous occupation of the slow channel A via the current through channel B. Sampling the number of charges Δn traversing the fast circuit during a fixed time interval Δt defines an approximate current $I = \Delta n / \Delta t$. When $\Gamma_{\alpha B}\Delta t$ is extremely small, the thus-defined current will be extremely noisy and will not yield reliable information on the instantaneous occupation of the slow channel. When in contrast the sampling time interval is very large such that $\Gamma_{\alpha A}\Delta t \gg 1$, the detector averages over several periods of filled and empty channel B, such that it is also not suitable as a detector of charge fluctuations. Only in the intermediate regime $\Gamma_{\alpha B}\Delta t \gg 1$ and $\Gamma_{\alpha A}\Delta t \ll 1$ may one use channel B as a detector of the instantaneous state of channel A. The trajectories one obtains in this case are similar to a quantum point contact (QPC) monitoring nearby charge fluctuations. In fact, the model can be used as a toy model for a QPC [5], and its simplicity also allows one to determine the QPC fidelity.

5.3.2 Thermodynamic Interpretation

The system obviously obeys three conservation laws: the energy exchange through all four junctions is globally conserved $I_E^{(LA)} + I_E^{(RA)} + I_E^{(LB)} + I_E^{(RB)} = 0$, and each circuit obeys matter conservation $I_M^{(LA)} + I_M^{(RA)} = 0$ and $I_M^{(LB)} + I_M^{(RB)} = 0$. Thus, the long-term entropy production can be specified by counting in general five out of eight energy and matter currents computable for the four-terminal system. When however we also keep all temperatures equal, $\beta_{LA} = \beta_{RA} = \beta_{LB} = \beta_{RB} = \beta$, the entropy production becomes

$$\dot{S}_i = \beta(\mu_{LA} - \mu_{RA})I_M^{(LA)} + \beta(\mu_{LB} - \mu_{RB})I_M^{(LB)}, \tag{5.37}$$

and can thus be quantified using only the matter currents traveling from left to right through the circuits. This implies that one will no longer have a single-circuit fluctuation theorem. Instead, denoting the joint probability that n_A and n_B electrons have traveled through channels A and B at time t, respectively, by $P_{n_A,n_B}(t)$, the fluctuation theorem becomes

$$\lim_{t\to\infty} \frac{P_{+n_A,+n_B}(t)}{P_{-n_A,-n_B}(t)} = e^{\beta(\mu_{LA}-\mu_{RA})n_A + \beta(\mu_{LB}-\mu_{RB})n_B}. \tag{5.38}$$

If only one circuit is monitored, one will in general not find a fluctuation theorem.

5.3.3 Reduced Dynamics

A special case however occurs for time-scale separation [6–8]. For example, when the tunneling rates through channel B are much larger than those through channel A, it is possible to find a fluctuation theorem for channel A only,

$$\lim_{t\to\infty} \frac{P_{+n_A}(t)}{P_{-n_A}(t)} = e^{\mathscr{A}_{\text{red}}n_A}, \tag{5.39}$$

where however the affinity $\mathscr{A}_{\text{red}} \neq \beta(\mu_{LA} - \mu_{RA})$ is different from the unperturbed affinity. In particular, with regard to experimental verifications of the fluctuation theorem by electrometers that exchange energy with the system [9, 10], it is important to note that the detector backaction modifies the affinity [4, 11]. There are multiple possible ways of deriving such a reduced fluctuation theorem. For example, the corresponding limit can be obtained by performing the limit in the characteristic polynomial. Here, we instead aim at deriving a coarse-grained rate equation [12] accounting for the dynamics of circuit A only when B is very fast. The rate equation for the dot occupations $\dot{P}_{\rho\sigma} = \sum_{\rho'\sigma'} \mathscr{L}_{\rho\sigma,\rho'\sigma'} P_{\rho'\sigma'}$ can be formally written as a rate equation accounting only for the dynamics of

$P_\rho = \sum_\sigma P_{\rho\sigma}$ via

$$\dot{P}_\rho = \sum_{\rho'\sigma\sigma'} \mathscr{L}_{\rho\sigma,\rho'\sigma'} P_{\rho\sigma} = \sum_{\rho'} \left[\sum_{\sigma\sigma'} \mathscr{L}_{\rho\sigma,\rho'\sigma'} \frac{P_{\rho'\sigma'}}{P_{\rho'}} \right] P_{\rho'}$$

$$\overset{\text{scale separation}}{\longrightarrow} \sum_{\rho'} \left[\sum_{\sigma\sigma'} \mathscr{L}_{\rho\sigma,\rho'\sigma'} \frac{\bar{P}_{\rho'\sigma'}}{\bar{P}_{\rho'}} \right] P_{\rho'} = \mathscr{L}_{\rho\rho'} P_{\rho'}, \tag{5.40}$$

which just requires us to compute the conditional probability of the system of being in state $(\rho\sigma)$ given that channel A is in state ρ. Strictly speaking, the mapping to a reduced Markovian master equation only works in the limit where $\Gamma_{\alpha B}, \Gamma_{\alpha B}^U \to \infty$. Then, the conditioned rates become

$$\frac{P_{00}}{P_0} = \frac{\Gamma_{LB}}{\Gamma_{LB} + \Gamma_{RB}} (1 - f_{LB}) + \frac{\Gamma_{RB}}{\Gamma_{LB} + \Gamma_{RB}} (1 - f_{RB}),$$

$$\frac{P_{01}}{P_0} = \frac{\Gamma_{LB}}{\Gamma_{LB} + \Gamma_{RB}} f_{LB} + \frac{\Gamma_{RB}}{\Gamma_{LB} + \Gamma_{RB}} f_{RB},$$

$$\frac{P_{10}}{P_1} = \frac{\Gamma_{LB}^U}{\Gamma_{LB}^U + \Gamma_{RB}^U} (1 - f_{LB}^U) + \frac{\Gamma_{RB}^U}{\Gamma_{LB}^U + \Gamma_{RB}^U} (1 - f_{RB}^U),$$

$$\frac{P_{11}}{P_1} = \frac{\Gamma_{LB}^U}{\Gamma_{LB}^U + \Gamma_{RB}^U} f_{LB}^U + \frac{\Gamma_{RB}^U}{\Gamma_{LB}^U + \Gamma_{RB}^U} f_{RB}^U, \tag{5.41}$$

and thus just represent two non-equilibrium steady states of channel B, depending on whether channel A is empty or filled, respectively. The reduced rate equation for channel A,

$$\begin{pmatrix} \dot{P}_0 \\ \dot{P}_1 \end{pmatrix} = \begin{pmatrix} -\mathscr{L}_{10} & +\mathscr{L}_{01} \\ +\mathscr{L}_{10} & -\mathscr{L}_{01} \end{pmatrix} \begin{pmatrix} P_0 \\ P_1 \end{pmatrix}, \tag{5.42}$$

is determined by the transition rates

$$\mathscr{L}_{01} = \mathscr{L}_{00,10} \frac{\bar{P}_{10}}{\bar{P}_1} + \mathscr{L}_{01,10} \frac{\bar{P}_{11}}{\bar{P}_1}$$

$$= \left[\Gamma_{LA}^U (1 - f_{LA}^U) + \Gamma_{RA}^U (1 - f_{RA}^U) \right] \left[\frac{\Gamma_{LB}^U}{\Gamma_{LB}^U + \Gamma_{RB}^U} f_{LB}^U + \frac{\Gamma_{RB}^U}{\Gamma_{LB}^U + \Gamma_{RB}^U} f_{RB}^U \right]$$

$$+ \left[\Gamma_{LA} (1 - f_{LA}) + \Gamma_{RA} (1 - f_{RA}) \right]$$

$$\times \left[\frac{\Gamma_{LB}^U}{\Gamma_{LB}^U + \Gamma_{RB}^U} (1 - f_{LB}^U) + \frac{\Gamma_{RB}^U}{\Gamma_{LB}^U + \Gamma_{RB}^U} (1 - f_{RB}^U) \right], \tag{5.43}$$

$$\mathscr{L}_{10} = \mathscr{L}_{10,00}\frac{\bar{P}_{00}}{\bar{P}_0} + \mathscr{L}_{11,01}\frac{\bar{P}_{01}}{\bar{P}_0}$$

$$= [\Gamma_{LA}f_{LA} + \Gamma_{RA}f_{RA}]\left[\frac{\Gamma_{LB}}{\Gamma_{LB} + \Gamma_{RB}}(1 - f_{LB}) + \frac{\Gamma_{RB}}{\Gamma_{LB} + \Gamma_{RB}}(1 - f_{RB})\right]$$

$$+ [\Gamma_{LA}^U f_{LA}^U + \Gamma_{RA}^U f_{RA}^U]\left[\frac{\Gamma_{LB}}{\Gamma_{LB} + \Gamma_{RB}}f_{LB} + \frac{\Gamma_{RB}}{\Gamma_{LB} + \Gamma_{RB}}f_{RB}\right].$$

One can still identify the terms in the rate matrix responsible for electronic jumps to the left and right reservoirs, since $\mathscr{L}_{\rho\sigma,\rho'\sigma'} = \mathscr{L}^{(L)}_{\rho\sigma,\rho'\sigma'} + \mathscr{L}^{(R)}_{\rho\sigma,\rho'\sigma'}$. The rates for the left and right junctions of channel A,

$$\mathscr{L}^{(\alpha)}_{01} = \Gamma_{\alpha A}^U (1 - f_{\alpha A}^U)\left[\frac{\Gamma_{LB}^U}{\Gamma_{LB}^U + \Gamma_{RB}^U}f_{LB}^U + \frac{\Gamma_{RB}^U}{\Gamma_{LB}^U + \Gamma_{RB}^U}f_{RB}^U\right]$$

$$+ \Gamma_{\alpha A}(1 - f_{\alpha A})\left[\frac{\Gamma_{LB}^U}{\Gamma_{LB}^U + \Gamma_{RB}^U}(1 - f_{LB}^U) + \frac{\Gamma_{RB}^U}{\Gamma_{LB}^U + \Gamma_{RB}^U}(1 - f_{RB}^U)\right],$$

$$\mathscr{L}^{(\alpha)}_{10} = \Gamma_{\alpha A}f_{\alpha A}\left[\frac{\Gamma_{LB}}{\Gamma_{LB} + \Gamma_{RB}}(1 - f_{LB}) + \frac{\Gamma_{RB}}{\Gamma_{LB} + \Gamma_{RB}}(1 - f_{RB})\right]$$

$$+ \Gamma_{\alpha A}^U f_{\alpha A}^U\left[\frac{\Gamma_{LB}}{\Gamma_{LB} + \Gamma_{RB}}f_{LB} + \frac{\Gamma_{RB}}{\Gamma_{LB} + \Gamma_{RB}}f_{RB}\right],$$

$$(5.44)$$

however now no longer obey local detailed balance in general. Only when channel A is completely independent on the occupation of channel B, which formally corresponds to $\Gamma_{\alpha A}^U = \Gamma_{\alpha A}$ and $f_{\alpha A}^U = f_{\alpha A}$, do we recover the original local detailed balance relations for channel A. The fluctuation theorem for particle transfers through channel A (5.39) can now be directly deduced by introducing a particle counting field

$$\mathscr{L}(\chi) = \begin{pmatrix} -\mathscr{L}^{(L)}_{10} - \mathscr{L}^{(R)}_{10} & +\mathscr{L}^{(L)}_{01} + \mathscr{L}^{(R)}_{01}e^{+i\chi} \\ +\mathscr{L}^{(L)}_{10} + \mathscr{L}^{(R)}_{10}e^{-i\chi} & -\mathscr{L}^{(L)}_{01} - \mathscr{L}^{(R)}_{01} \end{pmatrix}, \qquad (5.45)$$

and the reduced affinity can be expressed in terms of the matrix elements of the reduced rate matrix

$$\mathscr{A}_{\text{red}} = \ln\left[\frac{\mathscr{L}^{(L)}_{10} \mathscr{L}^{(R)}_{01}}{\mathscr{L}^{(L)}_{01} \mathscr{L}^{(R)}_{10}}\right]. \qquad (5.46)$$

Exercise 5.5 (Reduced affinity) Confirm the validity of the reduced affinity in Eq. (5.46).

Depending on the parameters, this effective affinity may now even become negative, even though a positive bias is applied. This already demonstrates that

$n\mathscr{A}_{\text{red}}$ cannot quantify the entropy production of subsystem A, unless in the decoupled case $U = 0$, where the effective affinity reduces to the conventional one $\mathscr{A}_0 = \ln[f_{LA}(1 - f_{RA})/((1 - f_{LA})f_{RA})] = \beta(\mu_{LA} - \mu_{RA})$. Physically, a negative affinity for positive bias applied to channel A simply means that the average current through channel A will flow against the applied potential gradient. It can be shown by simple inspection that in the case where channel A has only flat tunneling rates $\Gamma^U_{\alpha A} = \Gamma_{\alpha A}$, the affinity cannot become negative $\mathscr{A}_{\text{red}} \geq 0$, such that to generate useful power in channel A one has to go beyond the wide-band approximation.

In experiments trying to verify the fluctuation theorem by full counting statistics, this is usually done by coupling two circuits: variations in the current through the fast circuit (e.g., a QPC) can be associated with quantum jumps in the slow circuit (e.g., a DQD), and the counting statistics of the slow circuit is reconstructed from this data [13, 14]. The energy exchange between system and detector however modifies the effective affinity of the slow circuit, and this correction must be taken into account when interpreting the experimental results. For the present simple toy model such an effect is also present, and indeed the toy model may be used for example to calculate detection errors and detector backaction explicitly [5].

5.3.4 Drag Current

The drag current may also be analyzed in greater detail. To reduce the number of parameters, we parametrize the energy dependence of the tunneling rates by just two dimensionless parameters δ and Δ:

$$
\begin{aligned}
\Gamma_{LA} &= \Gamma\frac{e^{+\delta}}{\cosh(\delta)}, & \Gamma^U_{LA} &= \Gamma\frac{e^{-\delta}}{\cosh(\delta)}, \\
\Gamma_{RA} &= \Gamma\frac{e^{-\delta}}{\cosh(\delta)}, & \Gamma^U_{RA} &= \Gamma\frac{e^{+\delta}}{\cosh(\delta)}, \\
\Gamma_{LB} &= \Gamma\frac{e^{+\Delta}}{\cosh(\Delta)}, & \Gamma^U_{LB} &= \Gamma\frac{e^{-\Delta}}{\cosh(\Delta)}, \\
\Gamma_{RB} &= \Gamma\frac{e^{-\Delta}}{\cosh(\Delta)}, & \Gamma^U_{RB} &= \Gamma\frac{e^{+\Delta}}{\cosh(\Delta)},
\end{aligned}
\tag{5.47}
$$

where $\delta = \Delta = 0$ reproduces the case of completely symmetric and equal tunneling rates Γ. In contrast, the limit $\delta \to \infty$ parametrizes a case where in channel A an electron can only enter and exit from the left at energy ε_A, whereas tunneling processes to the right are allowed at energy $\varepsilon_A + U$. Transport in this limit is thus only allowed by the exchange of energy with channel B. Similarly, the parameter Δ controls channel B.

Thermodynamically, since only the total entropy production must be positive $\dot{S}_{\text{i}} \geq 0$, it is possible that, e.g., $(\mu_{LA} - \mu_{RA})I^{(LA)}_M < 0$, which would imply that the current through circuit A may be directed against the bias. To preserve the second law, this requires that the current through circuit B must then be directed with

Fig. 5.8 *Contour plot of efficiency η versus dimensionless bias voltage βV$_A$ (horizontal axis) and βV$_B$ (vertical axis). Contour lines range* from $\eta = 0.0$ *to* $\eta = 1.0$ *in steps of* $\Delta\eta = 0.1$. *White regions* denote formally negative efficiencies, where the drag current is not directed against the bias. For finite asymmetry in the energy-dependent tunneling rates δ and Δ, the Coulomb drag can only drive the current against a finite bias. Other parameters have been chosen as $\varepsilon_A = (\mu_{LA} + \mu_{RA})/2 - U/2$, $\varepsilon_B = (\mu_{LB} + \mu_{RB})/2 - U/2$, $\beta U = 1$, and $\Delta = \delta = 10$

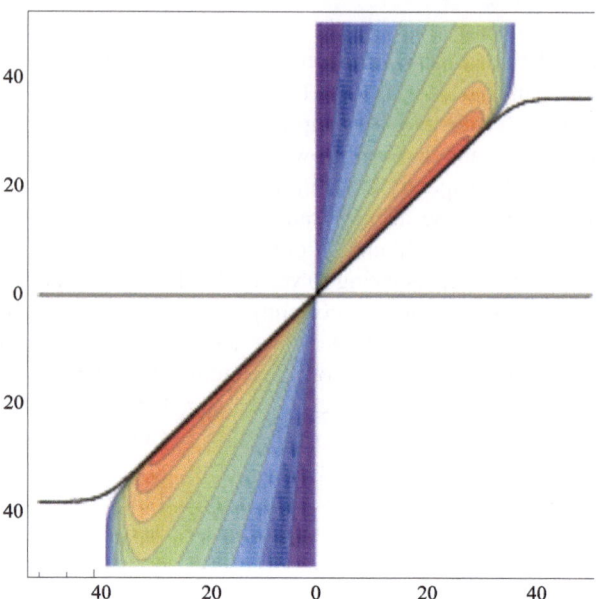

the bias. Furthermore, to obtain a positive total entropy production, we must have $I_M^{(LB)} V_B \geq -I_M^{(LA)} V_A$, where $V_{A/B} = \mu_{LA/B} - \mu_{RA/B}$ are the bias voltages of the corresponding circuits. This simply means that the power produced by channel A is smaller than the heat dissipated by channel B, and the current through B drags the current through A via its Coulomb interaction. This enables one to define the efficiency of this Coulomb drag engine [15, 16] as

$$\eta = -\frac{I_M^{(LA)} V_A}{I_M^{(LB)} V_B} = -\frac{I_M^{(LA)} \beta V_A}{I_M^{(LB)} \beta V_B} = -\frac{I_M^{(LA)} \ln \frac{f_{LA}(1-f_{RA})}{(1-f_{LA})f_{RA}}}{I_M^{(LB)} \ln \frac{f_{LB}(1-f_{RB})}{(1-f_{LB})f_{RB}}}, \qquad (5.48)$$

which is always upper bounded, $\eta \leq 1$. Note however that η may formally become negative (when, e.g., both currents flow with the bias anyway, but in these parameter regions it does not make sense to define this efficiency). Figure 5.8 displays the efficiency as a function of dimensionless bias voltages βV_A and βV_B for a specific case (see caption).

5.4 SET Monitored by a Low-Transparency QPC

A qualitatively different non-equilibrium situation is generated when, within a multicomponent reservoir, its constituents interact directly, i.e., even without the presence of the system. When this interaction is sufficiently weak, one may still assume a tensor product decomposition of the respective equilibrium states. To introduce some interesting dynamics, the interaction may be modified by the presence of the

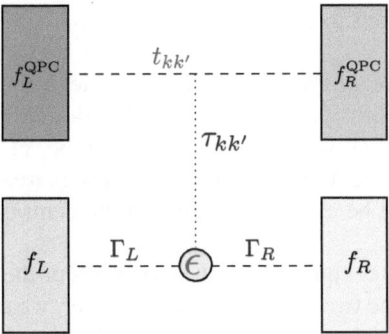

Fig. 5.9 Sketch of a low-transparency QPC (in fact, a two-component bath with the components held at different chemical potentials) monitoring an SET. The bare tunneling amplitudes $t_{kk'}$ through the QPC are modified to $t_{kk'} + \tau_{kk'}$ (*dotted line*) when the SET is occupied. *Dashed lines* denote tunneling processes treated perturbatively, and the *dotted line* denotes the exchange of energy between the SET circuit and the QPC. To lowest order in the interaction however, it turns out that the QPC does not have an effect on the SET dynamics

quantum system. A prototypical example for such a bath is a QPC: its leads are held at different potentials, but through a tiny contact charges may directly tunnel from one lead to another. The tunneling process is however highly sensitive to the presence of nearby charges, and in the Hamiltonian this is modeled by tunneling amplitudes that are changed when a nearby quantum dot is occupied. When the baseline tunneling amplitudes (in a low-transparency QPC) are small, we may apply the master equation formalism without great effort. We note that the backaction of realistic QPCs on the probed system can be quite small, as high-precision tests of counting statistics—performed with a QPC coupled to an SET—have revealed an impressive validity of the master equation approach [17].

5.4.1 Model

The Hamiltonian of the system depicted in Fig. 5.9 reads

$$H_{\mathrm{S}} = \varepsilon d^{\dagger} d,$$

$$H_{\mathrm{B}} = \sum_k \varepsilon_{kL} c_{kL}^{\dagger} c_{kL} + \sum_k \varepsilon_{kL} c_{kR}^{\dagger} c_{kR} + \sum_k \varepsilon_{kL} \gamma_{kL}^{\dagger} \gamma_{kL} + \sum_k \varepsilon_{kL} \gamma_{kR}^{\dagger} \gamma_{kR}, \quad (5.49)$$

$$H_{\mathrm{I}} = \left[\sum_k t_{kL} d c_{kL}^{\dagger} + \sum_k t_{kR} d c_{kR}^{\dagger} + \mathrm{h.c.} \right] + \left[\sum_{kk'} (t_{kk'} + d^{\dagger} d \tau_{kk'}) \gamma_{kL} \gamma_{k'R}^{\dagger} + \mathrm{h.c.} \right],$$

where ε denotes the dot level, $c_{k\alpha}$ are annihilation operators for electrons on SET lead α, and $\gamma_{k\alpha}$ are the annihilation operators for the QPC lead α. The QPC baseline tunneling amplitude is given by $t_{kk'}$ and describes the scattering of an electron from mode k in the left lead to mode k' in the right QPC contact. When the nearby SET is occupied, it is modified to $t_{kk'} + \tau_{kk'}$, where $\tau_{kk'}$ represents the change of the tunneling amplitude.

We note that the QPC Hamiltonian commutes with the dot Hamiltonian $\varepsilon d^\dagger d$ of the SET circuit. Thus, if we treat only the SET dot coupled to the QPC circuit, we would recover a model of the pure dephasing type, cf. Sect. 3.1. This already highlights that when the central dot of the SET only is treated as a system, neither energy nor particles will be exchanged between the central dot of the SET and the QPC in this model.

We will derive a master equation for the dynamics of the SET due to the interaction with the QPC and the two SET contacts. However, we are interested not only in the charge counting statistics of the SET but also in the counting statistics through the QPC. Since the quantum jumps of the QPC cannot be identified a posteriori in the SET rates, we will derive them microscopically. The Liouvillian for the SET-contact interaction is well known and has been stated previously (we insert particle counting fields at the right lead):

$$\mathscr{L}_{\mathrm{SET}}(\chi) = \begin{pmatrix} -\Gamma_L f_L - \Gamma_R f_R & +\Gamma_L(1-f_L) + \Gamma_R(1-f_R)e^{+\mathrm{i}\chi} \\ +\Gamma_L f_L + \Gamma_R f_R e^{-\mathrm{i}\chi} & -\Gamma_L(1-f_L) - \Gamma_R(1-f_R) \end{pmatrix}. \quad (5.50)$$

We will therefore derive the dissipator for the SET-QPC interaction separately, tracking the tunneled QPC electrons. To keep track of the tunneled QPC electrons, we insert a virtual detector operator in the respective tunneling Hamiltonian:

$$\begin{aligned}
H_{\mathrm{I}}^{\mathrm{QPC}} &= \sum_{kk'} \left(t_{kk'}\mathbf{1} + d^\dagger d \tau_{kk'} \right) B^\dagger \gamma_{kL} \gamma_{k'R}^\dagger + \sum_{kk'} \left(t_{kk'}^*\mathbf{1} + d^\dagger d \tau_{kk'}^* \right) B \gamma_{k'R} \gamma_{kL}^\dagger \\
&= \mathbf{1} \otimes B^\dagger \otimes \sum_{kk'} t_{kk'} \gamma_{kL} \gamma_{k'R}^\dagger + \mathbf{1} \otimes B \otimes \sum_{kk'} t_{kk'}^* \gamma_{k'R} \gamma_{kL}^\dagger \\
&\quad + d^\dagger d \otimes B^\dagger \otimes \sum_{kk'} \tau_{kk'} \gamma_{kL} \gamma_{k'R}^\dagger + d^\dagger d \otimes B \otimes \sum_{kk'} \tau_{kk'}^* \gamma_{k'R} \gamma_{kL}^\dagger. \quad (5.51)
\end{aligned}$$

Note that we have implicitly performed the mapping to a tensor product representation of the fermionic operators, which is unproblematic here as between SET and QPC no particle exchange takes place, and the electrons in the QPC and the SET may be treated as different particle types. To simplify the system, we assume that the change of tunneling amplitudes affects all modes in the same manner, i.e., $\tau_{kk'} = \tilde{\tau} t_{kk'}$, where τ represents the extent to which the tunneling is modified when an electron is present in the SET. This enables us to combine some coupling operators,

$$\begin{aligned}
H_{\mathrm{I}}^{\mathrm{QPC}} &= \left[\mathbf{1} + \tilde{\tau} d^\dagger d \right] \otimes B^\dagger \otimes \sum_{kk'} t_{kk'} \gamma_{kL} \gamma_{k'R}^\dagger \\
&\quad + \left[\mathbf{1} + \tilde{\tau}^* d^\dagger d \right] \otimes B \otimes \sum_{kk'} t_{kk'}^* \gamma_{k'R} \gamma_{kL}^\dagger. \quad (5.52)
\end{aligned}$$

The evident advantage of this approximation is that only two correlation functions have to be computed. Furthermore, we note that we treat the QPC perturbatively in

$t_{kk'}$ and in τ. We can now straightforwardly (since the baseline tunneling term is not included in the bath Hamiltonian) map to the interaction picture:

$$B_1(\tau) = \sum_{kk'} t_{kk'} \gamma_{kL} \gamma_{k'R}^\dagger e^{-i(\varepsilon_{kL} - \varepsilon_{k'R})\tau},$$

$$B_2(\tau) = \sum_{kk'} t_{kk'}^* \gamma_{k'R} \gamma_{kL}^\dagger e^{+i(\varepsilon_{kL} - \varepsilon_{k'R})\tau}. \tag{5.53}$$

For the first bath correlation function we obtain

$$C_{12}(\tau) = \sum_{kk'} \sum_{\ell\ell'} t_{kk'} t_{\ell\ell'}^* e^{-i(\varepsilon_{kL} - \varepsilon_{k'R})\tau} \langle \gamma_{kL} \gamma_{k'R}^\dagger \gamma_{\ell'R} \gamma_{\ell L}^\dagger \rangle$$

$$= \sum_{kk'} |t_{kk'}|^2 e^{-i(\varepsilon_{kL} - \varepsilon_{k'R})\tau} [1 - f_L(\varepsilon_{kL})] f_R(\varepsilon_{k'R})$$

$$= \frac{1}{2\pi} \iint T(\omega, \omega') [1 - f_L(\omega)] f_R(\omega') e^{-i(\omega-\omega')\tau} \, d\omega \, d\omega', \tag{5.54}$$

where we have introduced $T(\omega, \omega') = 2\pi \sum_{kk'} |t_{kk'}|^2 \delta(\omega - \varepsilon_{kL}) \delta(\omega - \varepsilon_{k'R})$. Note that in contrast to previous tunneling rates, this quantity is dimensionless. The integral factorizes when $T(\omega, \omega')$ factorizes (or when it is flat $T(\omega, \omega') = T_0$). In this case, the correlation function $C_{12}(\tau)$ is expressed as a product in the time domain, such that its Fourier transform will be given by a convolution integral,

$$\gamma_{12}(\Omega) = \int C_{12}(\tau) e^{+i\Omega\tau} \, d\tau = T_0 \int d\omega \, d\omega' \, [1 - f_L(\omega)] f_R(\omega') \delta(\omega - \omega' - \Omega)$$

$$= T_0 \int [1 - f_L(\omega)] f_R(\omega - \Omega) \, d\omega. \tag{5.55}$$

For the other correlation function, we have

$$\gamma_{21}(\Omega) = T_0 \int f_L(\omega) [1 - f_R(\omega + \Omega)] \, d\omega. \tag{5.56}$$

These correlation functions allow for a simple interpretation of the corresponding tunneling processes. For example, $\Gamma_{21}(\Omega)$ describes all tunneling processes from the left to the right lead, where the electron gains energy Ω from the system.

Exercise 5.6 (Correlation functions for the QPC) Show the validity of Eq. (5.56).

The structure of the Fermi functions demonstrates that the shift Ω can be included in the chemical potentials. Therefore, the technical problem is mapped to integrals of the type

$$I = \int f_1(\omega) [1 - f_2(\omega)] \, d\omega \tag{5.57}$$

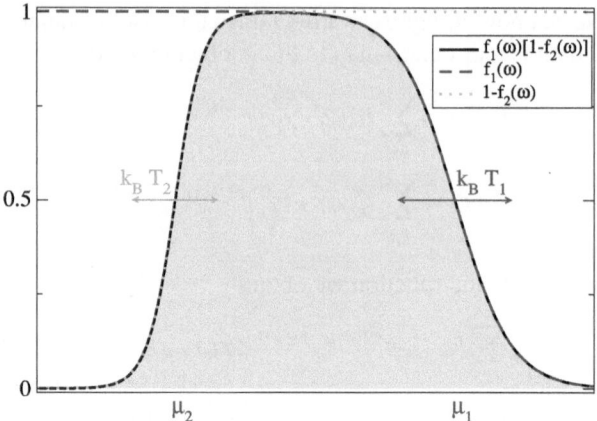

Fig. 5.10 Integrand in Eq. (5.57). As the Fermi function $f_1(\omega)$ (*red dashed curve*) declines from 1 to 0 at μ_1 and the factor $1 - f_2(\omega)$ (*orange dotted curve*) rises from 0 to 1 at μ_2, the product of both (*solid black*) will have support between μ_2 and μ_1, when $\mu_2 < \mu_1$. This becomes exact at zero temperature, where we obtain a product of two step functions and the area under the curve is given by the difference $\mu_1 - \mu_2$ as soon as $\mu_1 > \mu_2$ (and zero otherwise)

with Fermi functions $f_i(\omega)$. The zero temperature limit is particularly simple to solve: then, the Fermi functions become step functions, and the integral should behave as $I \approx (\mu_1 - \mu_2)\Theta(\mu_1 - \mu_2)$, where $\Theta(x)$ denotes the Heaviside Θ function; see also Fig. 5.10. For finite and different temperatures, the value of the integral can also be calculated. For simplicity however we constrain ourselves to the (experimentally relevant) case of equal but finite temperatures ($\beta_1 = \beta_2 = \beta$) and different chemical potentials. In this case, we obtain

$$I = \int \frac{1}{(e^{\beta(\mu_2-\omega)} + 1)(e^{-\beta(\mu_1-\omega)} + 1)} \, d\omega$$

$$= \lim_{\delta \to \infty} \int \frac{1}{(e^{\beta(\mu_2-\omega)} + 1)(e^{-\beta(\mu_1-\omega)} + 1)} \frac{\delta^2}{\delta^2 + \omega^2} \, d\omega, \qquad (5.58)$$

where we have introduced the Lorentzian-shaped regulator in order to apply residue formulae later on. The poles of the integrand are given by

$$\omega_{\pm}^* = \pm i\delta,$$

$$\omega_{1,n}^* = \mu_1 + \frac{\pi}{\beta}(2n + 1) \qquad (5.59)$$

$$\omega_{2,n}^* = \mu_2 + \frac{\pi}{\beta}(2n + 1),$$

where $n \in \{0, \pm 1, \pm 2, \pm 3, \ldots\}$. Here, the first line results from the regularization and all remaining poles are given by the Matsubara frequencies. We can solve the integral by using the residue theorem; see also Fig. 5.11 for the integration contour.

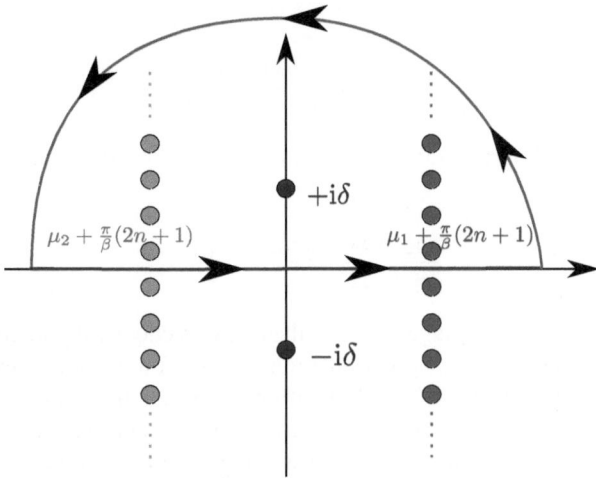

Fig. 5.11 Poles and integration contour for Eq. (5.57) in the complex plane. The integral along the *real axis* (*blue line*) is closed by an *arc* (*red curve*) in the upper complex plane, along which (due to the regulator) the integrand vanishes sufficiently fast. The sum over the residues at the Matsubara frequencies (*circles*) converges and—in the case of equal temperatures left and right—leads to a surprisingly simple result

Finally, we obtain for the integral

$$
\begin{aligned}
I = 2\pi i \lim_{\delta \to \infty} \Big\{ &\operatorname{Res} f_1(\omega)\big[1 - f_2(\omega)\big] \frac{\delta^2}{\delta^2 + \omega^2}\Big|_{\omega = +i\delta} \\
&+ \sum_{n=0}^{\infty} \operatorname{Res} f_1(\omega)\big[1 - f_2(\omega)\big] \frac{\delta^2}{\delta^2 + \omega^2}\Big|_{\omega = \mu_1 + \frac{\pi}{\beta}(2n+1)} \\
&+ \sum_{n=0}^{\infty} \operatorname{Res} f_1(\omega)\big[1 - f_2(\omega)\big] \frac{\delta^2}{\delta^2 + \omega^2}\Big|_{\omega = \mu_2 + \frac{\pi}{\beta}(2n+1)} \Big\} \\
= \frac{\mu_1 - \mu_2}{1 - e^{-\beta(\mu_1 - \mu_2)}}, & \qquad\qquad\qquad\qquad\qquad\qquad (5.60)
\end{aligned}
$$

which automatically obeys the simple zero temperature ($\beta \to \infty$) limit discussed before. With the replacements $\mu_1 \to \mu_R + \Omega$ and $\mu_2 \to \mu_L$, we obtain for the Fourier transform of the first bath correlation function

$$
\gamma_{12}(\Omega) = T_0 \frac{\Omega - V}{1 - e^{-\beta(\Omega - V)}}, \qquad\qquad\qquad\qquad (5.61)
$$

where $V = \mu_L - \mu_R$ is the QPC bias voltage. Likewise, with the replacements $\mu_1 \to \mu_L$ and $\mu_2 \to \mu_R - \Omega$, the Fourier transform of the second bath correlation function becomes

$$\gamma_{21}(\Omega) = T_0 \frac{\Omega + V}{1 - e^{-\beta(\Omega+V)}}. \tag{5.62}$$

We note that these Fourier transforms obey the Kubo–Martin–Schwinger (KMS)-type relation

$$\frac{\gamma_{12}(+\Omega)}{\gamma_{21}(-\Omega)} = e^{+\beta(\Omega-V)}, \tag{5.63}$$

where now the QPC bias voltage assumes the role of a chemical potential. Naturally, without an applied QPC bias, QPC transitions from left to right are just as probable as transitions from right to left, and at infinite bias, the transport becomes unidirectional. Now we can calculate the transition rates in our system (containing the virtual detector and the quantum dot) for a nondegenerate system spectrum. Since now the detector is part of our system, its state is not only characterized by the number of charges on the SET dot $a \in \{0, 1\}$ but also by the number of charges n that have tunneled through the QPC and have thereby changed the detector state:

$$\dot{\rho}_{(a,n)(a,n)} = \sum_{b,m} \gamma_{(a,n)(b,m),(a,n)(b,m)} \rho_{(b,m)(b,m)}$$

$$- \left[\sum_{b,m} \gamma_{(b,m)(a,n),(b,m)(a,n)} \right] \rho_{(a,n)(a,n)}. \tag{5.64}$$

Shortening the notation by omitting the double indices, we may also write

$$\dot{\rho}_{aa}^{(n)} = \sum_{b,m} \gamma_{(a,n),(b,m)} \rho_{bb}^{(m)} - \left[\sum_{b,m} \gamma_{(b,m),(a,n)} \right] \rho_{aa}^{(n)}, \tag{5.65}$$

where $\rho_{aa}^{(n)} = \rho_{(a,n),(a,n)}$ and $\gamma_{(a,n),(b,m)} = \gamma_{(a,n)(a,n),(b,m)(b,m)}$. It is evident that the coupling operators $A_1 = (1 + \tilde{\tau} d^\dagger d) \otimes B^\dagger$ and $A_2 = (1 + \tilde{\tau}^* d^\dagger d) \otimes B$ only allow for sequential tunneling through the QPC at lowest order (i.e., $m = n \pm 1$) and do not induce transitions between different dot states (i.e., $a = b$), such that the only nonvanishing contributions may arise for

$$\gamma_{(0,n)(0,n+1)} = \gamma_{12}(0)\langle 0, n|A_2|0, n+1\rangle\langle 0, n|A_1^\dagger|0, n+1\rangle^* = \gamma_{12}(0),$$

$$\gamma_{(0,n)(0,n-1)} = \gamma_{21}(0)\langle 0, n|A_1|0, n-1\rangle\langle 0, n|A_2^\dagger|0, n-1\rangle^* = \gamma_{21}(0),$$

$$\gamma_{(1,n)(1,n+1)} = \gamma_{12}(0)\langle 1, n|A_2|1, n+1\rangle\langle 1, n|A_1^\dagger|1, n+1\rangle^* = \gamma_{12}(0)|1 + \tilde{\tau}|^2,$$

$$\gamma_{(1,n)(1,n-1)} = \gamma_{21}(0)\langle 1, n|A_1|1, n-1\rangle\langle 1, n|A_2^\dagger|1, n-1\rangle^* = \gamma_{21}(0)|1 + \tilde{\tau}|^2. \tag{5.66}$$

The remaining terms just account for the normalization.

Exercise 5.7 (Normalization terms) Compute the remaining rates,

$$\sum_m \gamma_{(0,m)(0,m),(0,n)(0,n)} \quad \text{and} \quad \sum_m \gamma_{(1,m)(1,m),(1,n)(1,n)},$$

explicitly.

Adopting the notation of conditional master equations, this leads to the connected system

$$\dot{\rho}_{00}^{(n)} = \gamma_{12}(0)\rho_{00}^{(n+1)} + \gamma_{21}(0)\rho_{00}^{(n-1)} - [\gamma_{12}(0) + \gamma_{21}(0)]\rho_{00}^{(n)},$$

$$\dot{\rho}_{11}^{(n)} = |1 + \tilde{\tau}|^2 \gamma_{12}(0)\rho_{11}^{(n+1)} + |1 + \tilde{\tau}|^2 \gamma_{21}(0)\rho_{11}^{(n-1)} \qquad (5.67)$$

$$- |1 + \tilde{\tau}|^2 [\gamma_{12}(0) + \gamma_{21}(0)]\rho_{11}^{(n)},$$

such that after Fourier transformation with the counting field ξ for the QPC, we obtain the following dissipator:

$$\mathscr{L}_{\mathrm{QPC}}(\xi) = \begin{pmatrix} [\gamma_{21}(e^{+i\xi} - 1) + \gamma_{12}(e^{-i\xi} - 1)] & 0 \\ 0 & |1 + \tilde{\tau}|^2 [\gamma_{21}(e^{+i\xi} - 1) + \gamma_{12}(e^{-i\xi} - 1)] \end{pmatrix}.$$
$$(5.68)$$

The unusual occurrence of counting fields on the diagonal of the Liouvillian results from the fact that the QPC transitions happen while the SET circuit is stationary. Obviously, the counting statistics of the QPC could not have been deduced directly from the Liouvillian describing the SET dynamics alone: the QPC does not have any effect on the SET dynamics (which is of course an artifact of our perturbative treatment) as can be seen by setting $\xi \to 0$. More closely analyzing the Fourier transforms of the bath correlation functions,

$$\gamma_{21} = \gamma_{21}(0) = T_0 \frac{V}{1 - e^{-\beta V}},$$

$$\gamma_{12} = \gamma_{12}(0) = T_0 \frac{V}{e^{+\beta V} - 1}, \qquad (5.69)$$

we see that for sufficiently large QPC bias voltages, transport becomes unidirectional and only one contribution remains while the other one is exponentially suppressed. The sum of both Liouvillians (5.50) and (5.68) constitutes the total dissipator:

$$\mathscr{L}(\chi, \xi) = \mathscr{L}_{\mathrm{SET}}(\chi) + \mathscr{L}_{\mathrm{QPC}}(\xi), \qquad (5.70)$$

which can be used to calculate the probability distributions for tunneling through both transport channels (QPC and SET).

Exercise 5.8 (QPC current) Show that the stationary state of the SET is unaffected by the additional QPC dissipator and calculate the stationary current through the QPC for Liouvillian (5.70).

5.4.2 Detector Limit

When we consider the case $\{\Gamma_L, \Gamma_R\} \ll \{T_0 V, |1 + \tilde{\tau}| T_0 V\}$, we approach a bistable system, and the counting statistics approaches the case of telegraph noise.

When the dot is empty or filled throughout, respectively, the QPC current can easily be determined as

$$I_0 = [\gamma_{21}(0) - \gamma_{12}(0)], \qquad I_1 = |1 + \tilde{\tau}|^2 [\gamma_{21}(0) - \gamma_{12}(0)]. \qquad (5.71)$$

Since the moment-generating function in this case just corresponds to two counter-propagating Poisson processes, the noise $S = \frac{d}{dt}(\langle n^2 \rangle - \langle n \rangle^2)$ becomes

$$S_0 = [\gamma_{21}(0) + \gamma_{12}(0)], \qquad S_1 = |1 + \tilde{\tau}|^2 [\gamma_{21}(0) + \gamma_{12}(0)]. \qquad (5.72)$$

For finite time intervals Δt, the number of electrons Δn tunneling through the QPC during the time interval $[t, t + \Delta t]$ is determined by the probability distribution

$$P_{\Delta n}(\Delta t) = \frac{1}{2\pi} \int_{-\pi}^{+\pi} \text{Tr}\left\{ e^{\mathcal{L}(0,\xi)\Delta t - i\Delta n\xi} \rho(t) \right\} d\xi, \qquad (5.73)$$

where $\rho(t)$ represents the initial density matrix at time t. This quantity can, e.g., be evaluated numerically. Generally, a periodic measurement of the charges that have traversed the QPC during time intervals Δt then maps to a fixed-point iteration for the density matrix.

When Δt is not too large (such that the stationary state is not really reached) and not too small (such that there are sufficiently many particles tunneling through the QPC to meaningfully define a current), the time-resolved QPC current can be extracted as follows. To simulate a measurement of the particle number, a random number $\alpha \in [0, 1]$ can be drawn and used to determine a certain particle number n. Technically, this is the number n for which $\sum_{i \leq n} P_{\Delta i}(\Delta t) \leq \alpha \leq \sum_{i \leq n+1} P_{\Delta i}(\Delta t)$. Measuring a certain particle number m corresponds to a projection, i.e., the system-detector density matrix is projected to a certain measurement outcome which occurs with the probability $P_{\Delta m}(\Delta t)$,

$$\rho = \sum_n \rho^{(n)} \otimes |n\rangle\langle n| \xrightarrow{m} \frac{\rho^{(m)}}{\text{Tr}\{\rho^{(m)}\}}. \qquad (5.74)$$

The density matrix after the measurement is then used as the initial state for the next iteration, and the ratio of measured particles divided by the measurement time gives a current estimate $I(t) \approx \frac{\Delta n}{\Delta t}$. Repeating this process, one obtains a trajectory as displayed in Fig. 5.12. Such current trajectories are used to track the full counting statistics of small systems, which is for a bimodal current however only reliable when the transport across the small system is unidirectional [13, 17]. The presented detector model naturally includes measurement errors for finite measurement times Δt. When Δt is too small, the measurement error will be large since there is a large uncertainty of associating the correct charge state to the measured current. When

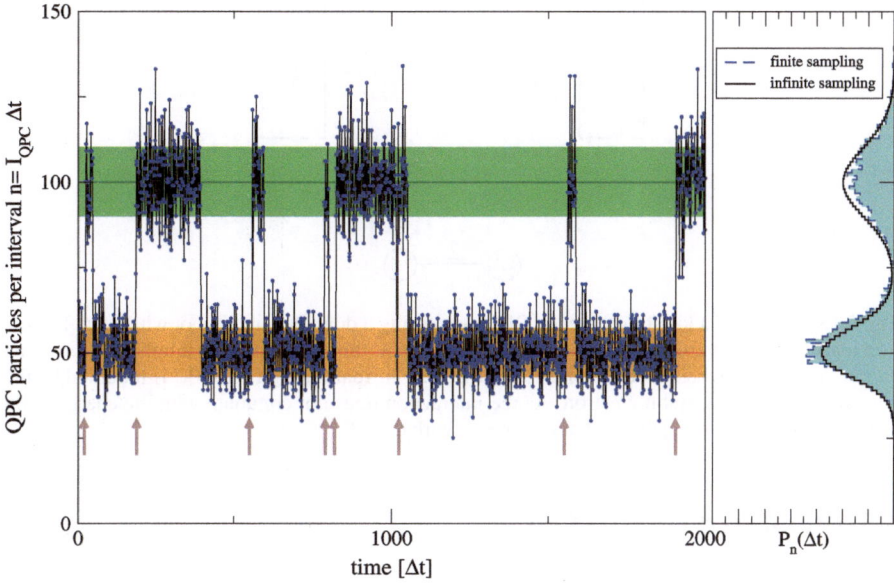

Fig. 5.12 Numerical simulation of the time-resolved QPC current for a slowly fluctuating SET dot occupation. At infinite SET bias, the QPC current allows one to reconstruct the full counting statistics of the SET, since each current blip (*arrows*) from low to high current—*horizontal lines* corresponding to Eq. (5.71) and *shaded regions* corresponding to the root of Eq. (5.72)—corresponds to an electron leaving the SET to its right junction. In this model, at finite SET bias, the direction of tunneling cannot be reliably deduced from the QPC current. Parameters: $\Gamma_L \Delta t = \Gamma_R \Delta t = 0.01$, $\gamma_{12}(0)\Delta t = |1 + \tilde{\tau}|^2 \gamma_{12}(0)\Delta t = 0$, $\gamma_{21}(0)\Delta t = 100.0$, $|1 + \tilde{\tau}|^2 \gamma_{21}(0)\Delta t = 50.0$, SET held at infinite bias ($f_L = 1$, $f_R = 0$). The *right panel* shows the corresponding probability distribution $P_n(\Delta t)$ versus $n = I\Delta t$, where the *shaded (dashed) curve* is sampled from the *left panel*, and the *black thick curve* is the theoretical limit Eq. (5.73) that one would obtain when sampling was performed for an infinitely long trajectory, i.e., for infinitely many intervals of finite Δt

Δt is too large, the measurement error will also be large since the detector averages over several cycles of loaded and unloaded SET states. In an intermediate regime however, one can expect the detector to yield reliable and meaningful results.

Unfortunately, the presented detector model completely neglects the detector backaction on the probed system, which makes it different from, e.g., the system discussed in Sect. 5.3. As a consequence, e.g., the fluctuation theorems for matter transfer through QPC and SET can be shown to hold independently; i.e., denoting the probability of having n charges traverse the circuit i by $P_n^i(t)$, one can prove the following relations:

$$\lim_{t \to \infty} \frac{P_{+n}^{\text{SET}}(t)}{P_{-n}^{\text{SET}}(t)} = e^{n\beta_{\text{SET}} V_{\text{SET}}}, \qquad \lim_{t \to \infty} \frac{P_{+n}^{\text{QPC}}(t)}{P_{-n}^{\text{QPC}}(t)} = e^{n\beta_{\text{QPC}} V_{\text{QPC}}}. \qquad (5.75)$$

Exercise 5.9 (Independent fluctuation theorems) Confirm the validity of Eq. (5.75).

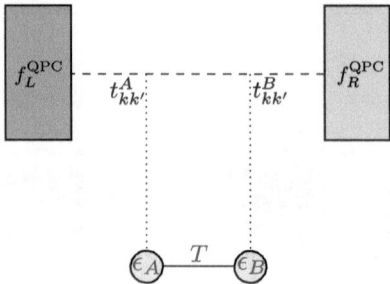

Fig. 5.13 Sketch of a low-transparency QPC monitoring a double quantum dot with a single electron loaded (charge qubit). When the electron is on dot A, the QPC tunneling amplitudes are given by $t^A_{kk'}$, and by $t^B_{kk'}$ when the electron is on dot B. The current through the QPC is modified by the position of the charge qubit electron. As the interaction does not commute with the charge qubit Hamiltonian, the charge qubit exchanges energy with the QPC

5.5 Monitored Charge Qubit

A low-transparency QPC may also be used to monitor a nearby charge qubit; see Fig. 5.13. The QPC performs a measurement of the electronic position, since its current is highly sensitive on it. When we identify the localized electronic states with eigenstates of the σ^z Pauli matrix, the current measurement thus corresponds to a σ^z measurement performed on the qubit. In addition, the mere presence of a detector of course also leads to a backaction on the probed system. Here, we will derive a master equation for the qubit to quantify this backaction and show that the detector induces decoherence in the energy eigenbasis of the qubit.

5.5.1 Model

The Hamiltonian of the charge qubit is given by

$$H_{\mathrm{CQB}} = \varepsilon_A d^\dagger_A d_A + \varepsilon_B d^\dagger_B d_B + T\left(d_A d^\dagger_B + d_B d^\dagger_A\right), \tag{5.76}$$

where we can safely neglect Coulomb interaction, since to form a charge qubit, the number of electrons on this double quantum dot (DQD) is fixed to one. The matrix representation is therefore just two dimensional in the $|n_A, n_B\rangle \in \{|10\rangle, |01\rangle\}$ basis and can be expressed by Pauli matrices,

$$H_{\mathrm{CQB}} = \begin{pmatrix} \varepsilon_A & T \\ T & \varepsilon_B \end{pmatrix} = \frac{\varepsilon_A + \varepsilon_B}{2}\mathbf{1} + \frac{\varepsilon_A - \varepsilon_B}{2}\sigma^z + T\sigma^x$$

$$\equiv \varepsilon\mathbf{1} + \Delta\sigma^z + T\sigma^x, \tag{5.77}$$

i.e., we may identify $d_A^\dagger d_A = \frac{1}{2}(1 + \sigma^z)$ and $d_B^\dagger d_B = \frac{1}{2}(1 - \sigma^z)$. The tunneling part of the QPC Hamiltonian reads as

$$H_I = d_A^\dagger d_A \otimes \sum_{kk'} t_{kk'}^A \gamma_{kL} \gamma_{k'R}^\dagger + d_B^\dagger d_B \otimes \sum_{kk'} t_{kk'}^B \gamma_{kL} \gamma_{k'R}^\dagger + \text{h.c.},\qquad(5.78)$$

where $t_{kk'}^{A/B}$ represents the tunneling amplitudes when the electron is localized on dots A and B, respectively. It does not commute with the Hamiltonian of the charge qubit, except for the pure-dephasing limit $T = 0$. After representing the charge qubit in terms of Pauli matrices, the full Hamiltonian reads as

$$\begin{aligned}
H = {}& \varepsilon \mathbf{1} + \Delta \sigma^z + T \sigma^x \\
& + \frac{1}{2}[1 + \sigma^z] \otimes \left[\sum_{kk'} t_{kk'}^A \gamma_{kL} \gamma_{k'R}^\dagger + \sum_{kk'} t_{kk'}^{A*} \gamma_{k'R} \gamma_{kL}^\dagger \right] \\
& + \frac{1}{2}[1 - \sigma^z] \otimes \left[\sum_{kk'} t_{kk'}^B \gamma_{kL} \gamma_{k'R}^\dagger + \sum_{kk'} t_{kk'}^{B*} \gamma_{k'R} \gamma_{kL}^\dagger \right] \\
& + \sum_k \varepsilon_{kL} \gamma_{kL}^\dagger \gamma_{kL} + \sum_k \varepsilon_{kR} \gamma_{kR}^\dagger \gamma_{kR}.
\end{aligned}\qquad(5.79)$$

To reduce the number of correlation functions, we again assume that all tunneling amplitudes are modified equally, $t_{kk'}^A = \tilde{\tau}_A t_{kk'}$ and $t_{kk'}^B = \tilde{\tau}_B t_{kk'}$ with baseline tunneling amplitudes $t_{kk'}$ and real constants τ_A and τ_B. Then, only a single correlation function needs to be calculated:

$$H_I = \left[\frac{\tilde{\tau}_A}{2}(1 + \sigma^z) + \frac{\tilde{\tau}_B}{2}(1 - \sigma^z) \right] \otimes \left[\sum_{kk'} t_{kk'} \gamma_{kL} \gamma_{k'R}^\dagger + \sum_{kk'} t_{kk'}^* \gamma_{k'R} \gamma_{kL}^\dagger \right],$$
$$(5.80)$$

which becomes explicitly

$$\begin{aligned}
C(\tau) = {}& \left\langle \sum_{kk'\ell\ell'} \left[t_{kk'} \gamma_{kL} \gamma_{k'R}^\dagger e^{-i(\varepsilon_{kL} - \varepsilon_{k'R})\tau} + t_{kk'}^* \gamma_{k'R} \gamma_{kL}^\dagger e^{+i(\varepsilon_{kL} - \varepsilon_{k'R})\tau} \right] \right. \\
& \left. \times \left[t_{\ell\ell'} \gamma_{\ell L} \gamma_{\ell'R}^\dagger + t_{\ell\ell'}^* \gamma_{\ell'R} \gamma_{\ell L}^\dagger \right] \right\rangle \\
= {}& \sum_{kk'} |t_{kk'}|^2 \left[e^{-i(\varepsilon_{kL} - \varepsilon_{k'R})\tau} \left\langle \gamma_{kL} \gamma_{k'R}^\dagger \gamma_{k'R} \gamma_{kL}^\dagger \right\rangle \right. \\
& \left. + e^{+i(\varepsilon_{kL} - \varepsilon_{k'R})\tau} \left\langle \gamma_{k'R} \gamma_{kL}^\dagger \gamma_{kL} \gamma_{k'R}^\dagger \right\rangle \right] \\
= {}& \frac{1}{2\pi} \int d\omega \, d\omega' \, T(\omega, \omega') \left[e^{-i(\omega - \omega')\tau} [1 - f_L(\omega)] f_R(\omega') \right. \\
& \left. + e^{+i(\omega - \omega')\tau} f_L(\omega) [1 - f_R(\omega')] \right],
\end{aligned}\qquad(5.81)$$

where we have in the last step replaced the summation by a continuous integration with $T(\omega, \omega') = 2\pi \sum_{kk'} |t_{kk'}|^2 \delta(\omega - \varepsilon_{kL}) \delta(\omega' - \varepsilon_{k'R})$. We directly conclude for the Fourier transform of the bath correlation function

$$
\begin{aligned}
\gamma(\Omega) &= \int d\omega \, d\omega' \, T(\omega, \omega') \big[\delta(\Omega - \omega + \omega')[1 - f_L(\omega)] f_R(\omega') \\
&\quad + \delta(\Omega + \omega - \omega') f_L(\omega)[1 - f_R(\omega')]\big] \\
&= \int d\omega \big[T(\omega, \omega - \Omega)[1 - f_L(\omega)] f_R(\omega - \Omega) \\
&\quad + T(\omega, \omega + \Omega) f_L(\omega)[1 - f_R(\omega + \Omega)]\big].
\end{aligned}
\tag{5.82}
$$

In what follows, we will consider the wide-band limit $T(\omega, \omega') = 1$ (the weak coupling limit enters the $\tilde{\tau}_{A/B}$ parameters), such that we may directly use the result from the previous Sect. 5.4:

$$
\gamma(\Omega) = \frac{\Omega + V}{1 - e^{-\beta(\Omega+V)}} + \frac{\Omega - V}{1 - e^{-\beta(\Omega-V)}},
\tag{5.83}
$$

where $V = \mu_L - \mu_R$ denotes the bias voltage of the QPC. We note that the Fourier transform of the correlation function obeys at zero bias voltage $V = 0$ the KMS relation

$$
\frac{\gamma(+\Omega)}{\gamma(-\Omega)} = e^{\beta\Omega}.
\tag{5.84}
$$

Since we are not interested in the QPC counting statistics here, we need not introduce any counting fields. However, the qubit-QPC interaction Hamiltonian does not commute with the Hamiltonian describing the charge qubit, which implies that we can expect the detector to have some impact on the qubit dynamics. The derivation of the master equation in the system energy eigenbasis requires diagonalization of the system Hamiltonian first:

$$
\begin{aligned}
E_- &= \varepsilon - \sqrt{\Delta^2 + T^2}, & |-\rangle &= \frac{-(\Delta + \sqrt{\Delta^2 + T^2})|0\rangle + T|1\rangle}{\sqrt{T^2 + (\Delta - \sqrt{\Delta^2 + T^2})^2}}, \\[2mm]
E_+ &= \varepsilon + \sqrt{\Delta^2 + T^2}, & |+\rangle &= \frac{-(\Delta - \sqrt{\Delta^2 + T^2})|0\rangle + T|1\rangle}{\sqrt{T^2 + (\Delta + \sqrt{\Delta^2 + T^2})^2}}.
\end{aligned}
\tag{5.85}
$$

Exercise 5.10 (Diagonalization of a single-qubit Hamiltonian) Calculate eigenvalues and eigenvectors of the system Hamiltonian.

Following the quantum optical master equation (compare Definition 2.3) we obtain a Lindblad master equation for the qubit

$$\dot{\rho} = -i[H_S + H_{LS}, \rho]$$

$$+ \sum_{abcd} \gamma_{ab,cd} \left[|a\rangle\langle b| \rho \big(|c\rangle\langle d|\big)^\dagger - \frac{1}{2} \big\{ \big(|c\rangle\langle d|\big)^\dagger |a\rangle\langle b|, \rho \big\} \right], \qquad (5.86)$$

where the summation only goes over the two energy eigenstates and H_{LS} denotes the frequency renormalization due to the detector (Lamb shift). Since the two eigenvalues of our system are nondegenerate, the Lamb-shift Hamiltonian is diagonal in the system energy eigenbasis and does not affect the dynamics of the populations, which decouples according to the rate equation

$$\dot{\rho}_{aa} = \sum_b \gamma_{ab,ab} \rho_{bb} - \left[\sum_b \gamma_{ba,ba} \right] \rho_{aa} \qquad (5.87)$$

completely from the coherences. In particular, we have

$$\dot{\rho}_{--} = \gamma_{--,--}\rho_{--} + \gamma_{-+,-+}\rho_{++} - \gamma_{--,--}\rho_{--} - \gamma_{+-,+-}\rho_{--}$$

$$= \gamma_{-+,-+}\rho_{++} - \gamma_{+-,+-}\rho_{--}, \qquad (5.88)$$

$$\dot{\rho}_{++} = \gamma_{+-,+-}\rho_{--} - \gamma_{-+,-+}\rho_{++},$$

which when written as a matrix becomes

$$\begin{pmatrix} \dot{\rho}_{--} \\ \dot{\rho}_{++} \end{pmatrix} = \begin{pmatrix} -\gamma_{+-,+-} & +\gamma_{-+,-+} \\ +\gamma_{+-,+-} & -\gamma_{-+,-+} \end{pmatrix} \begin{pmatrix} \rho_{--} \\ \rho_{++} \end{pmatrix}. \qquad (5.89)$$

The required damping coefficients read as

$$\gamma_{-+,-+} = \gamma(E_+ - E_-)\big|\langle -|A|+\rangle\big|^2 = \gamma\big(+2\sqrt{\Delta^2 + T^2}\big)\frac{T^2}{4(\Delta^2 + T^2)}(\tilde{\tau}_A - \tilde{\tau}_B)^2,$$

$$\gamma_{+-,+-} = \gamma(E_- - E_+)\big|\langle +|A|-\rangle\big|^2 = \gamma\big(-2\sqrt{\Delta^2 + T^2}\big)\frac{T^2}{4(\Delta^2 + T^2)}(\tilde{\tau}_A - \tilde{\tau}_B)^2.$$

$$(5.90)$$

Exercise 5.11 (Qubit dissipation) Show the validity of Eqs. (5.90).

This shows that when the QPC current is not dependent on the qubit state $\tilde{\tau}_A = \tilde{\tau}_B$, the dissipation on the qubit vanishes completely, which is consistent with our initial interaction Hamiltonian. In addition, in the pure dephasing limit $T \to 0$, we do not have any dissipative backaction of the measurement device on the qubit. Equation (5.89) obviously also preserves the trace of the density matrix.

5.5.2 Thermalization and Decoherence

The stationary density matrix must satisfy the relation

$$\frac{\bar{\rho}_{++}}{\bar{\rho}_{--}} = \frac{\gamma_{+-,+-}}{\gamma_{-+,-+}} = \frac{\gamma(-2\sqrt{\Delta^2 + T^2})}{\gamma(+2\sqrt{\Delta^2 + T^2})}. \tag{5.91}$$

When the QPC bias voltage vanishes (at equilibrium), we have

$$\frac{\gamma(-2\sqrt{\Delta^2 + T^2})}{\gamma(+2\sqrt{\Delta^2 + T^2})} \rightarrow e^{-\beta 2\sqrt{\Delta^2 + T^2}} = e^{-\beta(E_+ - E_-)}, \tag{5.92}$$

i.e., the qubit thermalizes with the temperature of the QPC. When the QPC bias voltage is large, the qubit is driven away from this thermal state.

Exercise 5.12 (Strongly monitored qubit) Calculate the stationary qubit state for the QPC held at infinite bias $V \rightarrow \pm\infty$.

The evolution of coherences decouples from the diagonal elements of the density matrix. The hermiticity of the density matrix allows us to consider only one coherence

$$\dot{\rho}_{-+} = -i(E_- + \Delta E_- - E_+ - \Delta E_+)\rho_{-+}$$

$$+ \left[\gamma_{--,++} - \frac{1}{2}(\gamma_{--,--} + \gamma_{++,++} + \gamma_{-+,-+} + \gamma_{+-,+-}) \right]\rho_{-+}, \tag{5.93}$$

where $\Delta E_\pm$ corresponds to the energy renormalization due to the Lamb shift, which induces a frequency renormalization of the qubit. The real part of the above equation is responsible for the damping of the coherence, and its calculation requires the evaluation of all remaining nonvanishing damping coefficients

$$\gamma_{-+,-+} + \gamma_{+-,+-}$$

$$= \left[\gamma(+2\sqrt{\Delta^2 + T^2}) + \gamma(-2\sqrt{\Delta^2 + T^2}) \right] \frac{T^2}{4(T^2 + \Delta^2)}(\tilde{\tau}_A - \tilde{\tau}_B)^2, \tag{5.94}$$

$$\gamma_{--,--} + \gamma_{++,++} - 2\gamma_{--,++} = \gamma(0)\frac{\Delta^2}{T^2 + \Delta^2}(\tilde{\tau}_A - \tilde{\tau}_B)^2.$$

Using the decomposition of the damping, we may now calculate the decoherence rate,

$$\gamma = \frac{1}{2}(\gamma_{-+,-+} + \gamma_{+-,+-}) + \frac{1}{2}(\gamma_{--,--} + \gamma_{++,++} - 2\gamma_{--,++})$$

$$= \frac{(\tilde{\tau}_A - \tilde{\tau}_B)^2}{8}\frac{T^2}{\Delta^2 + T^2}\left\{ (V + 2\sqrt{\Delta^2 + T^2})\coth\left[\frac{\beta}{2}(V + 2\sqrt{\Delta^2 + T^2}) \right] \right.$$

$$+ \left(V - 2\sqrt{\Delta^2 + T^2} \right) \coth\left[\frac{\beta}{2}\left(V - 2\sqrt{\Delta^2 + T^2} \right) \right] \bigg\}$$

$$+ \frac{(\tilde{\tau}_A - \tilde{\tau}_B)^2}{2} \frac{\Delta^2}{\Delta^2 + T^2} V \coth\left[\frac{\beta V}{2} \right], \tag{5.95}$$

which vanishes as reasonably expected when we set $\tilde{\tau}_A = \tilde{\tau}_B$. Note that $x \coth(x) \geq 1$ not only proves positivity of the decoherence rate (i.e., the coherences always decay), but it also enables one to obtain a rough lower bound

$$\gamma \geq \frac{(\tilde{\tau}_A - \tilde{\tau}_B)^2}{2\beta} \frac{T^2 + 2\Delta^2}{T^2 + \Delta^2} \tag{5.96}$$

on the dephasing rate. This lower bound is approached when the QPC voltage is rather small. For large voltages $|V| \gg \sqrt{\Delta^2 + T^2}$, the dephasing rate is given by

$$\gamma \approx \frac{(\tilde{\tau}_A - \tilde{\tau}_B)^2}{2} \frac{T^2 + 2\Delta^2}{T^2 + \Delta^2} |V| \tag{5.97}$$

and thus is limited by the voltage rather than the temperature. Thus, in any case the decoherence rate increases with both temperature and bias voltage of the QPC.

To study the time-resolved counting statistics of a QPC reading out the occupation of a DQD, it would now be straightforward to combine the results from this section and those of Sect. 5.2 to yield a model where the detector exchanges energy with the system. However, using an SET instead as detector, we can use our knowledge of the exact solution to obtain a true non-equilibrium environment, as will be discussed in Sect. 5.6.

5.6 High-Transparency QPC

As a toy model for a QPC we consider as in Sect. 5.3 again a single-electron transistor (SET). However, this time we lift the constraint of treating it in a weak coupling approximation and instead use the exact solution presented in Sect. 3.2. See Fig. 5.14.

5.6.1 Model

We assume that the SET detects nearby charge fluctuations (e.g., those of the nearby DQD) by some capacitive interaction; i.e., the interaction between the SET and another electronic quantum system is given by $H_I = A d^\dagger d$, where $d^\dagger d$ denotes the occupation of the central SET dot and $A = A^\dagger$ is some operator of the nearby quantum dot structure. When the upper circuit in Fig. 5.14 is solved exactly for its non-equilibrium stationary state, it can be considered a true non-equilibrium bath [8].

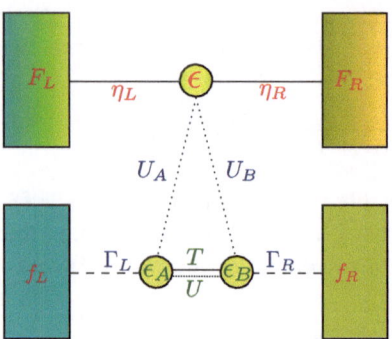

Fig. 5.14 Sketch of an SET circuit that is capacitively coupled via the additional Coulomb interactions U_A and U_B (*dotted lines*) to a nearby double quantum dot, which admits the exchange of energy. In contrast to Fig. 5.7, the *upper circuit* is treated non-perturbatively (*solid lines*) and thus constitutes a separate non-equilibrium environment. Despite the non-equilibrium nature of the bath, its correlation function depends in the long-term limit only on a single time argument

To obtain a simple master equation for the system, we first demand that the expectation value of the bath coupling operator vanish in the stationary state, which can be achieved by transforming $H_I \to A(d^\dagger d - \kappa \mathbf{1})$ and $H_S \to H_S + \kappa A$ with $\kappa = \lim_{t\to\infty}\langle d^\dagger(t)d(t)\rangle$. The exact solution of the QPC toy model only in Sect. 3.2 can be further simplified in the wide-band limit ($\delta_L \to \infty$ and $\delta_R \to \infty$). In this limit, we can easily compute the inverse Laplace transform of Eq. (3.38), and using $\eta = \eta_L + \eta_R > 0$ we can neglect all terms that decay for large times to obtain the asymptotic oscillatory solution (we relabel the SET tunneling rates from Sect. 3.2 $\Gamma \to \eta$ to avoid confusion):

$$d(t) \overset{t\to\infty}{\to} i\sum_k \left[\frac{e^{-i\varepsilon_{kL}t}t^*_{kL}c_{kL}}{-i\varepsilon_{kL}+i\varepsilon+\eta/2} + \frac{e^{-i\varepsilon_{kR}t}t^*_{kR}c_{kR}}{-i\varepsilon_{kR}+i\varepsilon+\eta/2}\right]$$
$$\equiv \sum_k \left[a_k(t)c_{kL} + b_k(t)c_{kR}\right], \tag{5.98}$$

where we have neglected all transient dynamics and thus only kept the asymptotic oscillatory behavior. To remain consistent, we will consider in the following only the wide-band limit.

To determine the shift κ, we calculate (also relabeling the SET Fermi functions from Sect. 3.2 $f_\alpha(\omega) \to F_\alpha(\omega)$ to avoid confusion)

$$\langle d^\dagger(t)d(t)\rangle \overset{t\to\infty}{\to} \sum_{kk'} \langle \left(a^*_k(t)c^\dagger_{kL} + b^*_k(t)c^\dagger_{kR}\right)\left(a_{k'}(t)c_{k'L} + b_{k'}(t)c_{k'R}\right)\rangle$$
$$= \sum_k \left[|a_k(t)|^2 F_L(\varepsilon_{kL}) + |b_k(t)|^2 F_R(\varepsilon_{kR})\right], \tag{5.99}$$

where we have used the fact that, in the Heisenberg picture, the expectation value must be computed with respect to the initial state, here assumed as a tensor product

$$\rho_0 = \frac{e^{-\beta_L(H_{\mathrm{B}}^{(L)} - \mu_L N_{\mathrm{B}}^{(L)})}}{\mathrm{Tr}\{e^{-\beta_L(H_{\mathrm{B}}^{(L)} - \mu_L N_{\mathrm{B}}^{(L)})}\}} \otimes \frac{e^{-\beta_R(H_{\mathrm{B}}^{(R)} - \mu_R N_{\mathrm{B}}^{(R)})}}{\mathrm{Tr}\{e^{-\beta_R(H_{\mathrm{B}}^{(R)} - \mu_R N_{\mathrm{B}}^{(R)})}\}} \tag{5.100}$$

of left and right SET leads. By introducing as usual the frequency-dependent tunneling rates

$$\eta_\alpha(\omega) = 2\pi \sum_k |t_{k\alpha}|^2 \delta(\omega - \varepsilon_{k\alpha}) \tag{5.101}$$

and using for consistency the wide-band approximation $\eta_\alpha(\omega) \approx \eta_\alpha$, one can convert the summation into integrals,

$$\kappa = \frac{1}{2\pi} \int d\omega \, \frac{\eta_L F_L(\omega) + \eta_R F_R(\omega)}{(\varepsilon - \omega)^2 + \eta^2/4}, \tag{5.102}$$

which can be solved exactly by the residue theorem, but we will be content with the above integral representation.

The correlation function constructed from the coupling operators (2.15) now generally depends on two time arguments, since we cannot assume that the reservoir is in an equilibrium state. After having applied the shift transformation in the interaction and system Hamiltonian, the correlation function reads

$$C(t_1, t_2) = \langle (d^\dagger(t_1)d(t_1) - \kappa \mathbf{1})(d^\dagger(t_2)d(t_2) - \kappa \mathbf{1})\rangle$$
$$\stackrel{t_i \to \infty}{\to} \langle d^\dagger(t_1)d(t_1)d^\dagger(t_2)d(t_2)\rangle - \kappa^2 \tag{5.103}$$

but can again be evaluated—since in the Heisenberg picture we have shifted the time dependence to the operators—with respect to the initial state (5.100).

We can therefore asymptotically (for large times t_1 and t_2) write the correlation function as

$$C(t_1, t_2) \stackrel{t_i \to \infty}{\to} \sum_{kk'qq'} \langle (a_k^*(t_1)c_{kL}^\dagger + b_k^*(t_1)c_{kR}^\dagger)(a_{k'}(t_1)c_{k'L} + b_{k'}(t_1)c_{k'R})$$

$$\times (a_q^*(t_2)c_{qL}^\dagger + b_q^*(t_2)c_{qR}^\dagger)(a_{q'}(t_2)c_{q'L} + b_{q'}(t_2)c_{q'R})\rangle - \kappa^2$$

$$= \sum_{kk'qq'} \big[a_k^*(t_1)a_{k'}(t_1)a_q^*(t_2)a_{q'}(t_2)\langle c_{kL}^\dagger c_{k'L}c_{qL}^\dagger c_{q'L}\rangle$$

$$+ a_k^*(t_1)a_{k'}(t_1)b_q^*(t_2)b_{q'}(t_2)\langle c_{kL}^\dagger c_{k'L}c_{qR}^\dagger c_{q'R}\rangle$$

$$+ a_k^*(t_1)b_{k'}(t_1)b_q^*(t_2)a_{q'}(t_2)\langle c_{kL}^\dagger c_{k'R}c_{qR}^\dagger c_{q'L}\rangle$$

$$+ b_k^*(t_1)a_{k'}(t_1)a_q^*(t_2)b_{q'}(t_2)\langle c_{kR}^\dagger c_{k'L}c_{qL}^\dagger c_{q'R}\rangle$$

$$+ b_k^*(t_1)b_{k'}(t_1)a_q^*(t_2)a_{q'}(t_2)\langle c_{kR}^\dagger c_{k'R}c_{qL}^\dagger c_{q'L}\rangle$$

$$+ b_k^*(t_1)b_{k'}(t_1)b_q^*(t_2)b_{q'}(t_2)\langle c_{kR}^\dagger c_{k'R}c_{qR}^\dagger c_{q'R}\rangle] - \kappa^2, \qquad (5.104)$$

where we have already used the fact that the initial state is a tensor product of thermal equilibrium states in the left and right leads of the SET to eliminate the expectation values that vanish most obviously. Furthermore, note that in the first and the last of the remaining terms above, there are two possible contractions of $\{k, k', q, q'\}$ into two pairs of indices and a single contribution where all indices are equal. The other terms only admit a single contraction into two pairs. The correlation function therefore becomes

$$C(t_1, t_2) = \sum_{kq}[|a_k(t_1)|^2|b_q(t_2)|^2 F_L(\varepsilon_{kL})F_R(\varepsilon_{qR})$$

$$+ a_k^*(t_1)a_k(t_2)b_q^*(t_2)b_q(t_1)F_L(\varepsilon_{kL})\big[1 - F_R(\varepsilon_{qR})\big]$$

$$+ b_k^*(t_1)b_k(t_2)a_q^*(t_2)a_q(t_1)F_R(\varepsilon_{kR})\big[1 - F_L(\varepsilon_{qL})\big]$$

$$+ |b_k(t_1)|^2|a_q(t_2)|^2 F_R(\varepsilon_{kR})F_L(\varepsilon_{qL})\big]$$

$$+ \sum_{k\neq q}[|a_k(t_1)|^2|a_q(t_2)|^2 F_L(\varepsilon_{kL})F_L(\varepsilon_{qL})$$

$$+ a_k^*(t_1)a_k(t_2)a_q^*(t_2)a_q(t_1)F_L(\varepsilon_{kL})\big[1 - F_L(\varepsilon_{qL})\big]$$

$$+ |b_k(t_1)|^2|b_q(t_2)|^2 F_R(\varepsilon_{kR})F_R(\varepsilon_{qR})$$

$$+ b_k^*(t_1)b_k(t_2)b_q^*(t_2)b_q(t_1)F_R(\varepsilon_{kR})\big[1 - F_R(\varepsilon_{qR})\big]]$$

$$+ \sum_{k}[|a_k(t_1)|^2|a_k(t_2)|^2 F_L(\varepsilon_{kL}) + |b_k(t_1)|^2|b_k(t_2)|^2 F_R(\varepsilon_{kR})] - \kappa^2,$$

$$(5.105)$$

where $F_\alpha(\omega)$ denotes the Fermi function compatible with the initial occupation of SET lead α. Note that one can skip the constraint in the second summation when the third summation is omitted:

$$C(t_1, t_2) = \sum_{kq}[|a_k(t_1)|^2|b_q(t_2)|^2 F_L(\varepsilon_{kL})F_R(\varepsilon_{qR})$$

$$+ |b_k(t_1)|^2|a_q(t_2)|^2 F_R(\varepsilon_{kR})F_L(\varepsilon_{qL})$$

$$+ |a_k(t_1)|^2|a_q(t_2)|^2 F_L(\varepsilon_{kL})F_L(\varepsilon_{qL})$$

$$+ |b_k(t_1)|^2|b_q(t_2)|^2 F_R(\varepsilon_{kR})F_R(\varepsilon_{qR})$$

$$+ a_k^*(t_1)a_k(t_2)b_q^*(t_2)b_q(t_1)F_L(\varepsilon_{kL})\big[1 - F_R(\varepsilon_{qR})\big]$$

$$+ b_k^*(t_1)b_k(t_2)a_q^*(t_2)a_q(t_1)F_R(\varepsilon_{kR})\big[1 - F_L(\varepsilon_{qL})\big]$$

$$+ a_k^*(t_1) a_k(t_2) a_q^*(t_2) a_q(t_1) F_L(\varepsilon_{kL})\big[1 - F_L(\varepsilon_{qL})\big]$$

$$+ b_k^*(t_1) b_k(t_2) b_q^*(t_2) b_q(t_1) F_R(\varepsilon_{kR})\big[1 - F_R(\varepsilon_{qR})\big]\big] - \kappa^2. \quad (5.106)$$

Direct inspection of the resulting combinations of coefficients yields that, for large times, the correlation function actually only depends on the difference of the time arguments,

$$C(t_1, t_2) \overset{t_i \to \infty}{\to} C(t_1 - t_2). \quad (5.107)$$

This enables one to define their Fourier transform in the long-term limit as usual. We proceed by introducing a continuum of reservoir frequencies, i.e., formally inserting Eq. (5.101), which enables us to represent the asymptotic correlation function by integrals:

$$\begin{aligned}
C(t_1 - t_2) &= \frac{1}{(2\pi)^2} \int d\omega \int d\omega' \frac{1}{[(\varepsilon - \omega)^2 + \eta^2/4][(\varepsilon - \omega')^2 + \eta^2/4]} \\
&\quad \times \big[\eta_L \eta_R F_L(\omega) F_R(\omega') + \eta_R \eta_L F_R(\omega) F_L(\omega') \\
&\quad + \eta_L \eta_L F_L(\omega) F_L(\omega') + \eta_R \eta_R F_R(\omega) F_R(\omega') \\
&\quad + \eta_L \eta_R e^{+i(\omega - \omega')(t_1 - t_2)} F_L(\omega)\big[1 - F_R(\omega')\big] \\
&\quad + \eta_R \eta_L e^{+i(\omega - \omega')(t_1 - t_2)} F_R(\omega)\big[1 - F_L(\omega')\big] \\
&\quad + \eta_L \eta_L e^{+i(\omega - \omega')(t_1 - t_2)} F_L(\omega)\big[1 - F_L(\omega')\big] \\
&\quad + \eta_R \eta_R e^{+i(\omega - \omega')(t_1 - t_2)} F_R(\omega)\big[1 - F_R(\omega')\big]\big] - \kappa^2 \\
&= \frac{1}{(2\pi)^2} \int d\omega \int d\omega' \frac{1}{[(\varepsilon - \omega)^2 + \eta^2/4][(\varepsilon - \omega')^2 + \eta^2/4]} \\
&\quad \times \big[+\eta_L \eta_R e^{+i(\omega - \omega')(t_1 - t_2)} F_L(\omega)\big[1 - F_R(\omega')\big] \\
&\quad + \eta_R \eta_L e^{+i(\omega - \omega')(t_1 - t_2)} F_R(\omega)\big[1 - F_L(\omega')\big] \\
&\quad + \eta_L \eta_L e^{+i(\omega - \omega')(t_1 - t_2)} F_L(\omega)\big[1 - F_L(\omega')\big] \\
&\quad + \eta_R \eta_R e^{+i(\omega - \omega')(t_1 - t_2)} F_R(\omega)\big[1 - F_R(\omega')\big]\big]. \quad (5.108)
\end{aligned}$$

In the last equation, we have used Eq. (5.102). This shift is therefore essential to yield a decay of the correlation function for large times, i.e., to motivate a Markovian approach. We note for consistency that for equal temperatures $\beta_L = \beta_R = \beta$ and vanishing chemical potentials $\mu_L = \mu_R = 0$ it is straightforward to demonstrate—using symmetry considerations—validity of the KMS condition (2.51), i.e., that $C(\tau) = C(-\tau - i\beta)$. Furthermore, one can solve all involved integrals exactly by summing up residues. This yields a representation in terms of hypergeometric functions (not shown for brevity).

It is much more convenient to represent the Fourier transform of the correlation function in terms of a convolution integral,

$$
\begin{aligned}
\gamma(\Omega) &= \int C(\tau) e^{+i\Omega\tau} \, d\tau \\
&= \frac{1}{2\pi} \int d\omega \, \frac{1}{[(\varepsilon - \omega)^2 + \eta^2/4][(\varepsilon - \omega - \Omega)^2 + \eta^2/4]} \\
&\quad \times \big[+\eta_L \eta_R F_L(\omega)\big[1 - F_R(\omega + \Omega)\big] + \eta_R \eta_L F_R(\omega)\big[1 - F_L(\omega + \Omega)\big] \\
&\quad + \eta_L \eta_L F_L(\omega)\big[1 - F_L(\omega + \Omega)\big] \\
&\quad + \eta_R \eta_R F_R(\omega)\big[1 - F_R(\omega + \Omega)\big]\big].
\end{aligned}
\tag{5.109}
$$

Here, the KMS condition $\gamma(-\Omega) = e^{-\beta\Omega}\gamma(+\Omega)$ is valid in equilibrium when $F_L(\omega) = F_R(\omega)$, which can be easily demonstrated by using $F(\omega)[1 - F(\omega+\Omega)] = e^{\beta\Omega} F(\omega + \Omega)[1 - F(\omega)]$. If an infinite bias is applied across the SET ($F_L(\omega) \to 1$, $F_R(\omega) \to 0$), one obtains simple values for the shift factor and the correlation function:

$$
\begin{aligned}
\kappa &\to \frac{\eta_L}{\eta_L + \eta_R}, \\
\gamma(\Omega) &\to \frac{\eta_L \eta_R}{\eta_L + \eta_R} \frac{2}{(\eta_L + \eta_R)^2 + \Omega^2}.
\end{aligned}
\tag{5.110}
$$

Exercise 5.13 (Correlation function at infinite bias) Confirm the validity of Eq. (5.110) in the infinite bias limit of the SET.

Furthermore, for very large η, we can neglect the dependence of the prefactor in the integrand on ω and Ω, such that the integrals reduce to the type discussed in Eq. (5.57). In any case, analysis in the complex plane would allow one to solve the above convolution integrals analytically in the wide-band limit.

Making everything explicit for a DQD in the lower circuit,

$$
H_S = \varepsilon_A d_A^\dagger d_A + \varepsilon_B d_B^\dagger d_B + T\big(d_A d_B^\dagger + d_B d_A^\dagger\big) + U d_A^\dagger d_A d_B^\dagger d_B, \tag{5.111}
$$

with the capacitive interaction between the DQD and the SET,

$$
H_I^{\mathrm{QPC}} = A \otimes d^\dagger d, \qquad A = U_A d_A^\dagger d_A + U_B d_B^\dagger d_B, \tag{5.112}
$$

implies that we will obtain a simple rate equation description in the renormalized system energy eigenbasis defined by the Hamiltonian

$$
\begin{aligned}
H_S' &= (\varepsilon_A + \kappa U_A) d_A^\dagger d_A + (\varepsilon_B + \kappa U_B) d_B^\dagger d_B \\
&\quad + T\big(d_A d_B^\dagger + d_B d_A^\dagger\big) + U d_A^\dagger d_A d_B^\dagger d_B
\end{aligned}
\tag{5.113}
$$

and the shift (5.102). The dependence of the shift κ on temperatures and chemical potentials of the SET circuit now has the consequence that the pointer basis (i.e., the basis within which one has a rate equation description for the density matrix dynamics) now depends on the parameters of the non-equilibrium reservoir. We may directly use the results from Sect. 5.2 to state the dissipators resulting from the coupling of the DQD with its leads. These are simply given by the Liouvillian in Eq. (5.20) and the rates in Eqs. (5.21) and (5.22), and all the properties discussed there also apply here. We note however, that due to the shift transformation, the original system parameters will be renormalized. In addition, the coupling to the SET circuit will also induce transitions between eigenstates of equal charge $|-\rangle$ and $|+\rangle$, which occur with the rates

$$\gamma_{-+,-+} = \gamma(E_+ - E_-)|\langle -|A|+\rangle|^2,$$
$$\gamma_{+-,+-} = \gamma(E_- - E_+)|\langle -|A|+\rangle|^2. \tag{5.114}$$

The ratio of these rates obeys local detailed balance when the SET circuit is in equilibrium, $F_L(\omega) = F_R(\omega)$:

$$\frac{\gamma_{-+,-+}}{\gamma_{+-,+-}} = \frac{\gamma(E_+ - E_-)}{\gamma(E_- - E_+)} \rightarrow e^{+\beta(E_+ - E_-)}. \tag{5.115}$$

This relation also holds when the SET circuit is at a finite chemical potential. However, since the SET circuit does not exchange particles with the DQD, the chemical potential does not show up in the local detailed balance relation at equilibrium. In contrast, when the SET is held at infinite bias, the above ratio becomes just one. The remaining matrix element can be readily computed,

$$|\langle -|A|+\rangle|^2 = |\langle -|U_A d_A^\dagger d_A + U_B d_B^\dagger d_B|+\rangle|^2$$
$$= \frac{T^2(U_A - U_B)^2}{4T^2 + (\varepsilon_L + \kappa U_A - \varepsilon_R - \kappa U_B)^2}, \tag{5.116}$$

where as a consistency check we note that at equal couplings $U_A = U_B$ the SET-associated transition rates vanish.

When we set the electronic tunneling rates through the DQD to zero, we recover in the subspace of the singly charged sector the situation of a charge qubit that is monitored by an SET. In contrast to the previous section however, here the SET is treated non-perturbatively.

5.6.2 Detector Backaction

Since the SET can be seen as a single-charge detector for the DQD state, its resulting additional dissipator can be used to estimate the effect of the detector presence

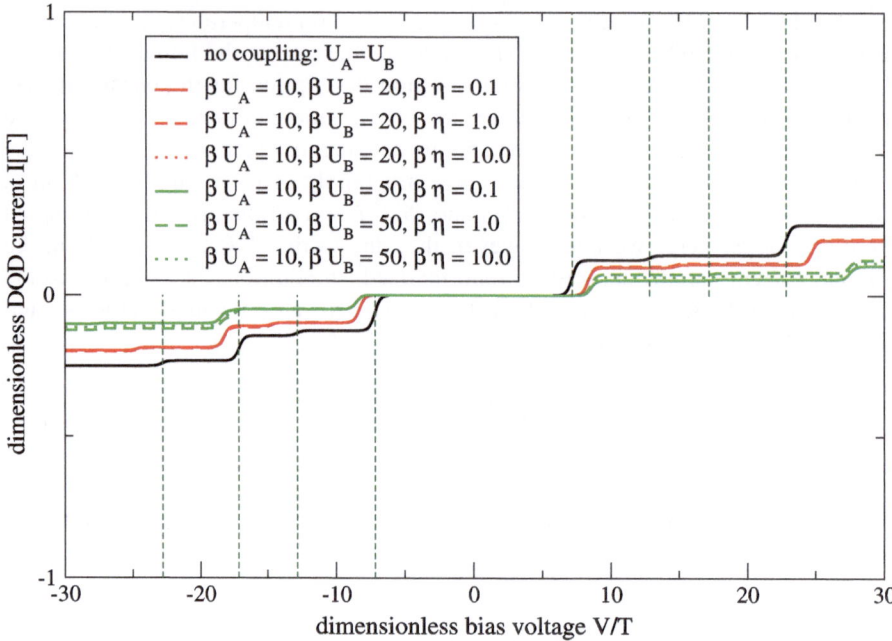

Fig. 5.15 Current through the DQD for different interaction strengths with the SET. At symmetric couplings $U_A = U_B$, the unperturbed DQD current (compare black solid curve in Fig. 5.6) is reproduced. For asymmetric capacitive interactions, the main effect of the SET is a shift of the transition frequencies (compare the shifted steps in *solid black, red,* and *light green curves*), whereas the actual coupling strength η between the two SET leads is of minor importance (*solid, dashed,* and *dotted curves*) for the DQD current. The DQD parameters have been chosen as in Fig. 5.6, and the SET was assumed at infinite bias, compare Eq. (5.110), with symmetric tunneling rates $\eta_L = \eta_R = 2\eta$

on the probed system. Here, it is not necessary to calculate the detector-induced decoherence: the detector presence already modifies the current through the DQD. Figure 5.15 shows the matter current through the DQD versus the DQD bias voltage at infinite SET bias and for varying SET tunneling rates (which go well beyond the weak coupling limit). Whereas the actual coupling strength of the SET quantum dot to its junctions is of minor importance, the shift of the system eigenfrequencies by Eq. (5.102) strongly modifies the current. It follows that when transport spectroscopy is performed by means of a high-transparency secondary circuit, one should keep in mind that the system properties may be modified by the presence of the detector.

Formally, we observe that the detector presence changes the transition rates of the system. Moreover, it also changes the pointer basis itself, within which the system assumes a rate equation dynamics. Finally, we note that, without information on the charges tunneling through the SET circuit, one will in general not obtain a fluctuation theorem for the charges and energies transferred through the DQD circuit only.

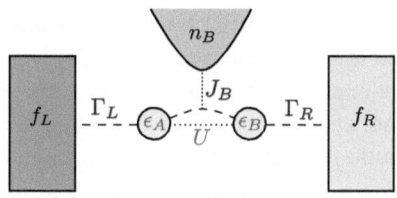

Fig. 5.16 Sketch of two quantum dots that are separately tunnel-coupled to their adjacent reservoir in the conventional way by rates Γ_L and Γ_R. The mere Coulomb interaction U only allows the exchange of energy between the dots, but with phonons present (*rounded terminals*), tunneling between A and B becomes possible (*dotted* and *dashed*). The device may act as a thermoelectric generator converting thermal gradients into power

5.7 Phonon-Assisted Tunneling

Many transport experiments are performed through molecules, where electrons may excite vibrational modes (phonons). In the bulk, coupling to phonons is also a relevant correction [18], and is also well accessible (in the weak coupling limit) by a master equation approach. Following existing works [4, 19] we consider two quantum dots, each tunnel-coupled to a separate fermionic bath. However, in the model, tunneling between the dots is only possible via the assistance of phonons; see Fig. 5.16.

5.7.1 Model

The system is described by the Hamiltonian

$$H_S = \varepsilon_A d_A^\dagger d_A + \varepsilon_B d_B^\dagger d_B + U d_A^\dagger d_A d_B^\dagger d_B \qquad (5.117)$$

with on-site energies $\varepsilon_A < \varepsilon_B$ (this assumption is without loss of generality due to the symmetry of the problem) and Coulomb interaction U. Its energy eigenstates coincide with the localized basis $|n_A, n_B\rangle$ with the dot occupations $n_A, n_B \in \{0, 1\}$. This structure makes it particularly simple to derive a master equation in rate equation representation (2.47). The jumps between states are triggered by the electronic tunneling Hamiltonians and the electron–phonon interaction,

$$H_I = \sum_k \left(t_{kL} d_A c_{kL}^\dagger + t_{kL}^* c_{kL} d_A^\dagger\right) + \sum_k \left(t_{kR} d_B c_{kR}^\dagger + t_{kR}^* c_{kR} d_B^\dagger\right)$$

$$+ \left(d_A d_B^\dagger + d_B d_A^\dagger\right) \otimes \sum_q \left(h_q a_q + h_q^* a_q^\dagger\right), \qquad (5.118)$$

where $c_{k\alpha}$ are fermionic and a_q bosonic annihilation operators. The three reservoirs

$$H_B = \sum_k \varepsilon_{kL} c_{kL}^\dagger c_{kL} + \sum_k \varepsilon_{kR} c_{kR}^\dagger c_{kR} + \sum_q \omega_q a_q^\dagger q_q \qquad (5.119)$$

are assumed to remain in separate thermal equilibrium states, such that the reservoir density matrix is assumed to be a product of the single density matrices. This automatically implies that the expectation value of linear combinations of the coupling operators vanishes. In the weak coupling limit, the rate matrix will be additively decomposed into contributions resulting from the electronic (L, R) and bosonic (B) reservoirs, $\mathscr{L} = \mathscr{L}_L + \mathscr{L}_R + \mathscr{L}_B$. From our results with the SET in Sect. 5.3, we may readily reproduce the rates for the electronic jumps without phonons. Ordering the basis as $\rho_{00,00}$, $\rho_{10,10}$, $\rho_{01,01}$, and $\rho_{11,11}$ and using for simplicity the wide-band limit $\Gamma_\alpha(\omega) \approx \Gamma_\alpha$, these read

$$
\mathscr{L}_L = \Gamma_L \begin{pmatrix} -f_L(\varepsilon_A) & 1 - f_L(\varepsilon_A) & 0 & 0 \\ +f_L(\varepsilon_A) & -[1 - f_L(\varepsilon_A)] & 0 & 0 \\ 0 & 0 & -f_L(\varepsilon_A + U) & 1 - f_L(\varepsilon_A + U) \\ 0 & 0 & +f_L(\varepsilon_A + U) & -[1 - f_L(\varepsilon_A + U)] \end{pmatrix},
$$

(5.120)

$$
\mathscr{L}_R = \Gamma_R \begin{pmatrix} -f_R(\varepsilon_B) & 0 & 1 - f_R(\varepsilon_B) & 0 \\ 0 & -f_R(\varepsilon_B + U) & 0 & 1 - f_R(\varepsilon_B + U) \\ +f_R(\varepsilon_B) & 0 & -[1 - f_R(\varepsilon_B)] & 0 \\ 0 & +f_R(\varepsilon_B + U) & 0 & -[1 - f_R(\varepsilon_B + U)] \end{pmatrix},
$$

where the electronic tunneling rates are as usual obtained via (in the wide-band limit) $\Gamma_\alpha \approx \Gamma_\alpha(\omega) = 2\pi \sum_k |t_{k\alpha}|^2 \delta(\omega - \varepsilon_{k\alpha})$ from the microscopic tunneling amplitudes $t_{k\alpha}$. We note that the Fermi functions are evaluated at the energy difference of the jump to which they refer. Although energy may be transferred between the left and right junctions without the presence of phonons, it is not possible to transfer charges.

To transfer charges through the DQD, coupling to the phonons is necessary. The bath correlation function for the phonon reservoir reads as

$$
\begin{aligned}
C(\tau) &= \sum_{qq'} \langle (h_q a_q e^{-i\omega_q \tau} + h_q^* a_q^\dagger e^{+i\omega_q \tau})(h_{q'} a_{q'} + h_{q'}^* a_{q'}^\dagger) \rangle \\
&= \sum_q |h_q|^2 \big[e^{-i\omega_q \tau} [1 + n_B(\omega_q)] + e^{+i\omega_q \tau} n_B(\omega_q) \big] \\
&= \frac{1}{2\pi} \int_0^\infty d\omega \, J_B(\omega) \big[e^{-i\omega\tau} [1 + n_B(\omega)] + e^{+i\omega\tau} n_B(\omega) \big] \\
&= \frac{1}{2\pi} \int_{-\infty}^{+\infty} J_B(\omega) [1 + n_B(\omega)] e^{-i\omega\tau},
\end{aligned}
$$

(5.121)

where we have used the symmetries of the Bose–Einstein distribution,

$$
n_B(\omega) = \frac{1}{e^{\beta\omega} - 1} = -[1 + n_B(-\omega)],
$$

(5.122)

and defined the spectral coupling density for bosons with support on the complete real axis by analytic continuation,

$$J_B(\omega) = \begin{cases} 2\pi \sum_q |h_q|^2 \delta(\omega - \omega_q): & \omega \geq 0, \\ -J_B(-\omega): & \omega < 0. \end{cases}$$

The above representation of the correlation function directly allows us to extract its Fourier transform,

$$\gamma(\omega) = J_B(\omega)\big[1 + n_B(\omega)\big], \tag{5.123}$$

which motivates the analytic continuation of the spectral coupling density above. For consistency, we just note that the KMS condition (2.51) is obeyed. We may thus readily evaluate the rates due to the phonon reservoirs, i.e., we have

$$\mathscr{L}_B = J_B(\varepsilon_B - \varepsilon_A) \begin{pmatrix} 0 & 0 & 0 & 0 \\ 0 & -n_B(\varepsilon_B - \varepsilon_A) & 1 + n_B(\varepsilon_B - \varepsilon_A) & 0 \\ 0 & +n_B(\varepsilon_B - \varepsilon_A) & -[1 + n_B(\varepsilon_B - \varepsilon_A)] & 0 \\ 0 & 0 & 0 & 0 \end{pmatrix}. \tag{5.124}$$

The rate matrices in Eqs. (5.120) and (5.124) can be used to extract the full electron–phonon counting statistics after all jumps have been identified. We have a three-terminal system, where the phonon terminal only allows for the exchange of energy. Adding further bosonic reservoirs that couple identically is possible without substantial effort, but for our purposes it suffices to consider only three terminals. With the conservation laws on matter and energy currents, we can expect to find a fluctuation theorem using only three counting fields. For simplicity, we decide to count matter at the left junction (χ) and energy transfers at the left (ξ) and the bosonic (η) junctions. To keep the description compact, we introduce the abbreviations

$$
\begin{aligned}
n_B &= n_B(\varepsilon_B - \varepsilon_A), & f_L &= f_L(\varepsilon_A), & f_R &= f_R(\varepsilon_B), \\
\bar{f}_L &= f_L(\varepsilon_A + U), & \bar{f}_R &= f_R(\varepsilon_B + U), & J_B &= J_B(\varepsilon_B - \varepsilon_A).
\end{aligned}
\tag{5.125}
$$

With the counting fields, the complete system becomes

$$
\begin{aligned}
\mathscr{L} = \Gamma_L &\begin{pmatrix} -f_L & (1 - f_L)e^{-i\chi}e^{-i\varepsilon_A\xi} & 0 & 0 \\ +f_L e^{+i\chi}e^{+i\varepsilon_A\xi} & -[1 - f_L] & 0 & 0 \\ 0 & 0 & -\bar{f}_L & (1 - \bar{f}_L)e^{-i\chi}e^{-i(\varepsilon_A+U)\xi} \\ 0 & 0 & +\bar{f}_L e^{+i\chi}e^{+i(\varepsilon_A+U)\xi} & -(1 - \bar{f}_L) \end{pmatrix} \\[2ex]
+ J_B &\begin{pmatrix} 0 & 0 & 0 & 0 \\ 0 & -n_B & (1 + n_B)e^{-i(\varepsilon_B-\varepsilon_A)\eta} & 0 \\ 0 & +n_B e^{+i(\varepsilon_B-\varepsilon_A)\eta} & -(1 + n_B) & 0 \\ 0 & 0 & 0 & 0 \end{pmatrix}
\end{aligned}
$$

$$+ \Gamma_R \begin{pmatrix} -f_R & 0 & 1-f_R & 0 \\ 0 & -\bar{f}_R & 0 & 1-\bar{f}_R \\ +f_R & 0 & -(1-f_R) & 0 \\ 0 & +\bar{f}_R & 0 & -(1-\bar{f}_R) \end{pmatrix}. \tag{5.126}$$

5.7.2 Thermodynamic Interpretation

Using energy $I_E^L + I_E^R + I_E^B = 0$ and matter $I_M^L + I_M^R = 0$ conservation, we can eliminate all but the currents we have chosen to monitor, such that we can write for the entropy production at steady state

$$\begin{aligned} \dot{S}_{\mathrm{i}} &= -\beta_{\mathrm{ph}} I_E^B - \beta_L \big(I_E^L - \mu_L I_M^L \big) - \beta_R \big(I_E^R - \mu_R I_M^R \big) \\ &= -\beta_B I_E^B - \beta_L \big(I_E^L - \mu_L I_M^L \big) + \beta_R \big(I_E^L + I_E^B - \mu_R I_M^L \big) \\ &= (\beta_R - \beta_B) I_E^B + (\beta_R - \beta_L) I_E^L + (\beta_L \mu_L - \beta_R \mu_R) I_M^L, \end{aligned} \tag{5.127}$$

which has the characteristic affinity-flux form. The characteristic polynomial $\mathscr{D}(\chi, \xi, \eta) = |\mathscr{L}(\chi, \xi, \eta) - \lambda \mathbf{1}|$ of the rate matrix (5.126) obeys the symmetry with the same affinities,

$$\begin{aligned} \mathscr{D}(-\chi, -\xi, -\eta) = \mathscr{D}\big(&+\chi + \mathrm{i}(\mu_L \beta_L - \mu_R \beta_R), +\xi + \mathrm{i}(\beta_R - \beta_L), \\ &+\eta + \mathrm{i}(\beta_R - \beta_B) \big), \end{aligned} \tag{5.128}$$

which implies a fluctuation theorem for the combined energy-particle counting statistics. We denote the probability that energy $\Delta\varepsilon$ has been transferred from the boson bath to the system and energy ΔE from the left lead to the system, and ΔN particles have traversed the system from left to right after time t by $P_{\Delta\varepsilon, \Delta E, \Delta N}(t)$. Then, the corresponding probability distribution obeys the fluctuation theorem

$$\lim_{t \to \infty} \frac{P_{+\Delta\varepsilon, +\Delta E, +\Delta N}}{P_{-\Delta\varepsilon, -\Delta E, -\Delta N}} = e^{(\mu_L \beta_L - \mu_R \beta_R)\Delta N + (\beta_R - \beta_L)\Delta E + (\beta_R - \beta_B)\Delta\varepsilon}, \tag{5.129}$$

where in the exponent we again identify the entropy production associated to the corresponding trajectories.

As a special property of the considered model, we note that an electron that is effectively transferred from left to right through the DQD must have absorbed energy $\varepsilon_B - \varepsilon_A$ from the phonon heat bath. As a result, the electronic matter current and the phonon energy current are tightly coupled, $I_E^B = (\varepsilon_B - \varepsilon_A) I_M^L$. The entropy production can therefore also be written in terms of two fluxes only, e.g., energy and matter currents at the left lead,

$$\dot{S}_{\mathrm{i}} = (\beta_R - \beta_L) I_E^L + \big[(\beta_L \mu_L - \beta_R \mu_R) + (\beta_R - \beta_B)(\varepsilon_B - \varepsilon_A) \big] I_M^L, \tag{5.130}$$

such that two counting fields generally suffice to quantify the entropy production. Disregarding the phonon energy counting, the characteristic polynomial obeys the reduced symmetry

$$\mathcal{D}(-\chi, -\xi, 0) = \mathcal{D}\big(+\chi + i\big[(\mu_L\beta_L - \mu_R\beta_R) + (\beta_R - \beta_B)(\varepsilon_B - \varepsilon_A)\big],$$
$$+ \xi + i(\beta_R - \beta_L), 0\big), \tag{5.131}$$

which implies that the fluctuation theorem can be expressed with only electronic energy and matter transfers:

$$\lim_{t \to \infty} \frac{P_{+\Delta E, +\Delta N}}{P_{-\Delta E, -\Delta N}} = e^{[(\beta_L\mu_L - \beta_R\mu_R) + (\beta_R - \beta_B)]\Delta N + (\beta_R - \beta_L)\Delta E}. \tag{5.132}$$

Realistically, monitoring the complete energy and matter flows is impossible. Usually, it is just possible to monitor transfers of single charges [13, 14]. Furthermore, the electronic temperatures at the left and right leads are usually the same. Noting that the affinity associated to the electronic energy current at the left lead vanishes at equal left and right temperatures $\beta_L = \beta_R = \beta_{\text{el}}$ and relabeling $\beta_{\text{ph}} = \beta_B$, we conclude that then the long-term entropy production

$$\dot{S}_i \to \big[\beta_{\text{el}}(\mu_L - \mu_R) + (\beta_{\text{el}} - \beta_{\text{ph}})(\varepsilon_B - \varepsilon_A)\big]I_M^L \tag{5.133}$$

can be quantified only in terms of the electronic particle counting statistics. Consistently, the characteristic polynomial obeys for equal electronic temperatures the symmetry

$$\mathcal{D}(-\chi, 0, 0) = \mathcal{D}\big(+\chi + i\big[\beta_{\text{el}}(\mu_L - \mu_R) + (\beta_{\text{el}} - \beta_{\text{ph}})(\varepsilon_B - \varepsilon_A)\big], 0, 0\big), \tag{5.134}$$

which implies the fluctuation theorem

$$\lim_{t \to \infty} \frac{P_{+\Delta N}}{P_{-\Delta N}} = e^{[\beta_{\text{el}}(\mu_L - \mu_R) + (\beta_{\text{el}} - \beta_{\text{ph}})(\varepsilon_B - \varepsilon_A)]\Delta N}. \tag{5.135}$$

The affinity in the above exponent demonstrates that at vanishing bias $\mu_L = \mu_R$, the current will become positive when $(\beta_{\text{el}} - \beta_{\text{ph}})(\varepsilon_B - \varepsilon_A) > 0$ and negative otherwise. Without further calculations we may also conclude that the current vanishes when the bias voltage reaches

$$\big(\mu_L^* - \mu_R^*\big) = V^* = -\frac{(\beta_{\text{el}} - \beta_{\text{ph}})(\varepsilon_B - \varepsilon_A)}{\beta_{\text{el}}}$$
$$= -\left(1 - \frac{T_{\text{el}}}{T_{\text{ph}}}\right)(\varepsilon_B - \varepsilon_A). \tag{5.136}$$

Figure 5.17 displays the current as a function of the bias voltage for different electronic and phonon temperature configurations.

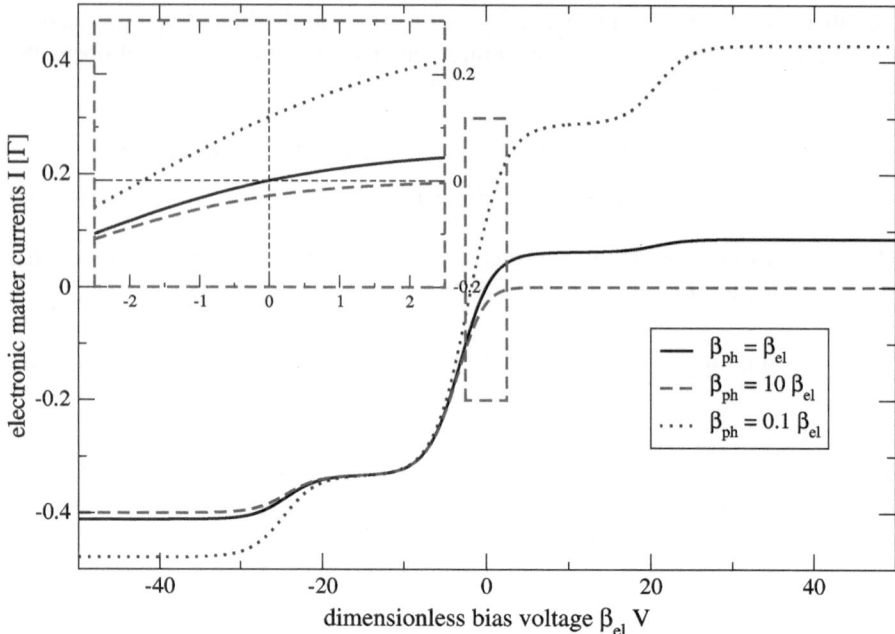

Fig. 5.17 Electronic matter current in units of $\Gamma_L = \Gamma_R = \Gamma$ versus dimensionless bias voltage $\beta_{el}V$. For low phonon temperatures $\beta_{ph}(\varepsilon_B - \varepsilon_A) \gg 1$, the current cannot flow from left to right, such that the system acts as a rectifier (*dashed red*). For large phonon temperatures $\beta_{ph}(\varepsilon_B - \varepsilon_A) \ll 1$, the energy driving the current against the bias (see *zoomed inset*) is supplied by the phonon bath. Other parameters: $\beta_{el}\varepsilon_B = 2$, $\beta_{el}\varepsilon_A = 0$, $\beta_{el}U = 10$, $J_B = \Gamma$, $\beta_L = \beta_R = \beta_{el}$, and $\mu_L = +V/2 = -\mu_R$ (Color figure online)

5.7.3 Thermoelectric Performance

We concentrate on the simple case discussed before and use $\beta_L = \beta_R = \beta_{el}$ and $\beta_{ph} = \beta_B$. In regions where the current runs against the bias, the power

$$P = -(\mu_L - \mu_R)I_M^L \tag{5.137}$$

becomes positive, and we can define an efficiency via

$$\eta = \frac{-(\mu_L - \mu_R)I_M^L}{\dot{Q}_{in}}, \tag{5.138}$$

where $\dot{Q}_{in}$ is the heat put into the system.

When the phonon temperature is larger than the electron temperature, the input heat is given by the positive energy flow from the hot phonon bath into the system, such that—due to the tight coupling property—the efficiency becomes trivially dependent on the bias voltage:

$$\eta_{T_{ph}>T_{el}} = \frac{P}{I_E^B} = -\frac{V}{\varepsilon_B - \varepsilon_A}. \tag{5.139}$$

At first sight, one might think that this efficiency could become larger than one. However, we should keep in mind that it is only valid in regimes where the power (5.137) is positive, which limits the applicability of these efficiencies to voltages within $V = 0$ and $V = V^*$ from Eq. (5.136). The maximum efficiency therefore becomes

$$\eta_{T_{ph}>T_{el}} < 1 - \frac{T_{el}}{T_{ph}}, \tag{5.140}$$

and is thus upper bounded by Carnot efficiency.

In the opposite case, the input heat is given by the sum of the energy currents entering from the hot electronic leads $\dot{Q}_{in} = \dot{Q}^L + \dot{Q}^R = I_E^L + I_E^R + P = -I_E^B + P$, such that the efficiency becomes

$$\eta_{T_{ph}<T_{el}} = \frac{P}{-I_E^B + P} = \frac{(\mu_L - \mu_R)}{(\varepsilon_B - \varepsilon_A) + (\mu_L - \mu_R)} = \frac{1}{1 + \frac{\varepsilon_B - \varepsilon_A}{\mu_L - \mu_R}}, \tag{5.141}$$

which also trivially depends on the bias voltage. Inserting the maximum bias voltage with positive power in Eq. (5.136), we obtain the maximum efficiency

$$\eta_{T_{ph}<T_{el}} < \frac{1}{1 + \frac{1}{\frac{T_{el}}{T_{ph}} - 1}} = 1 - \frac{T_{ph}}{T_{el}}, \tag{5.142}$$

which is also just the Carnot efficiency.

Unfortunately, Carnot efficiencies are reached at vanishing current, i.e., at zero power. At these parameters, a thermoelectric device is useless. It is therefore more practical to consider the efficiency at maximum power. However, since the currents depend in a highly nonlinear fashion on all parameters (coupling constants, temperatures, chemical potentials, and system parameters), this becomes a numerical optimization problem—unless one restricts the analysis to the linear response regime, where efficiency at maximum power is upper bounded [20] by the Curzon–Ahlborn [21] efficiency $\eta_P \leq 1 - \sqrt{\frac{T_{cold}}{T_{hot}}}$.

5.8 Beyond Weak Coupling: Phonon-Coupled Single Electron Transistor

As before, we consider a quantum dot model that is additionally coupled to phonons. To keep the analysis simple however, we follow Ref. [3] by considering an SET that is coupled to one, many, or even a continuum of phonon modes, as depicted in Fig. 5.18.

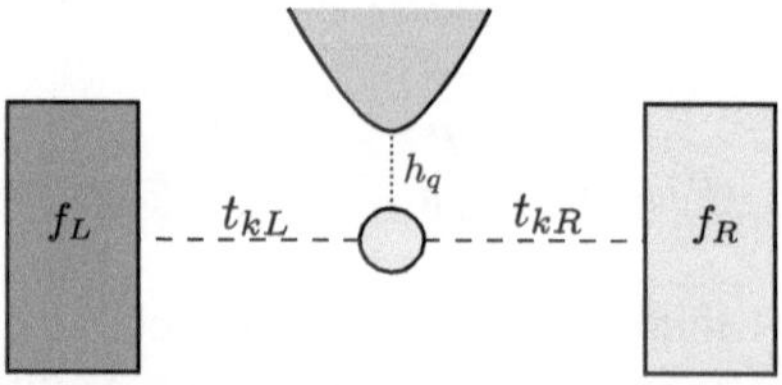

Fig. 5.18 Sketch of an SET that is capacitively coupled to a phonon reservoir. The interaction in the original Hamiltonian is of the pure dephasing type; i.e., the system energy will not be changed. A conventional master equation treatment would therefore yield no effect on the SET dynamics due to the phonon reservoir

5.8.1 Model

The SET Hamiltonian is as before given by

$$H_{\text{SET}} = \varepsilon d^\dagger d + \sum_{\alpha \in \{L,R\}} \sum_k [\varepsilon_{k\alpha} c_{k\alpha}^\dagger c_{k\alpha} + t_{k\alpha} d c_{k\alpha}^\dagger + t_{k\alpha}^* c_{k\alpha} d^\dagger]. \quad (5.143)$$

In addition however, the central dot of the SET now interacts

$$H_{\text{I}} = d^\dagger d \otimes \sum_{q=1}^{Q} [h_q a_q + h_q^* a_q^\dagger] \quad (5.144)$$

with a phonon reservoir $H_{\text{B}}^{\text{ph}} = \sum_q \omega_q a_q^\dagger a_q$ containing Q phonon modes. Obviously, the interaction commutes with the central dot part of the SET Hamiltonian. Therefore, if one would conventionally derive a master equation for the population dynamics of the central quantum dot, the additional phonon bath would not affect the populations of the central dot at all—the interaction is of pure dephasing type. In general however, this cannot be true: the interaction does not commute with the total SET Hamiltonian, and therefore one must expect the phonons to have some effect. Indeed, extensive calculations with only a single phonon mode whose dynamics is completely taken into account have revealed a strong suppression of the electronic current when strongly coupled phonons are present. This phenomenon has been termed the Franck–Condon blockade [22].

To treat such cases within a master equation approach, we apply a transformation to the full Hamiltonian $H' = U H U^\dagger$ with the unitary operator

$$U = \exp\left\{ d^\dagger d \sum_q \left(\frac{h_q^*}{\omega_q} a_q^\dagger - \frac{h_q}{\omega_q} a_q \right) \right\} \equiv e^{d^\dagger d A}. \quad (5.145)$$

The above transformation is known as the polaron or Lang–Firzov transformation [23, 24]. Obviously, the electronic leads are unaffected by the transformation, since $U c_{k\alpha} U^\dagger = c_{k\alpha}$, and also the central dot part is inert $U d^\dagger d U^\dagger = d^\dagger d$. There

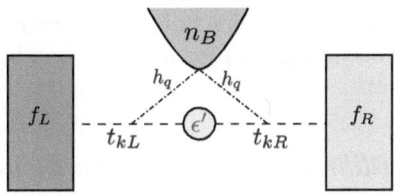

Fig. 5.19 After the polaron transformation, direct coupling between the central quantum dot and the phonons in Fig. 5.18 is transformed to the electronic tunnel couplings. The electron–phonon coupling may be treated non-perturbatively (*dash–dotted lines*) when the electronic tunnel couplings are treated perturbatively (*dashed lines*)

are multiple ways of proving the following relations:

$$U d U^\dagger = d e^{-A}, \qquad U d^\dagger U^\dagger = d^\dagger e^{+A},$$

$$U a_q U^\dagger = a_q - \frac{h_q^*}{\omega_q} d^\dagger d, \qquad U a_q^\dagger U^\dagger = a_q^\dagger - \frac{h_q}{\omega_q} d^\dagger d. \tag{5.146}$$

Exercise 5.14 (Polaron transformation) Show the validity of Eqs. (5.146).

These immediately also imply the relation

$$U a_q^\dagger a_q U^\dagger = a_q^\dagger a_q - \frac{d^\dagger d}{\omega_q}\left(h_q a_q + h_q^* a_q^\dagger\right) + \frac{|h_q|^2}{\omega_q^2} d^\dagger d. \tag{5.147}$$

After the polaron transformation, the Hamiltonian therefore reads

$$H' = \left(\varepsilon - \sum_q \frac{|h_q|^2}{\omega_q}\right) d^\dagger d + \sum_{k\alpha} \varepsilon_{k\alpha} c_{k\alpha}^\dagger c_{k\alpha} + \sum_q \omega_q a_q^\dagger a_q$$

$$+ \sum_{k\alpha}\left(t_{k\alpha} d c_{k\alpha}^\dagger e^{-A} + t_{k\alpha}^* c_{k\alpha} d^\dagger e^{+A}\right), \tag{5.148}$$

and thereby admits a new decomposition into system and bath Hamiltonians; see also Fig. 5.19. Most obviously, we observe a shift of the electronic level $\varepsilon \to \varepsilon' = \varepsilon - \sum_q \frac{|h_q|^2}{\omega_q}$. Second, the electronic tunneling terms between the central dot and the adjacent leads now become 'dressed' by exponential operators,

$$H_I' = \sum_{k\alpha}\left[t_{k\alpha} d c_{k\alpha}^\dagger e^{-\sum_q\left(\frac{h_q^*}{\omega_q} a_q^\dagger - \frac{h_q}{\omega_q} a_q\right)} + t_{k\alpha}^* c_{k\alpha} d^\dagger e^{+\sum_q\left(\frac{h_q^*}{\omega_q} a_q^\dagger - \frac{h_q}{\omega_q} a_q\right)}\right], \tag{5.149}$$

which demonstrates that every single electronic jump from the central dot to the leads may now trigger multiple phonon emissions or absorptions. This implies that a perturbative treatment in $t_{k\alpha}$ still enables a non-perturbative treatment of the phonon absorption and emission amplitudes h_q. Furthermore, this leads to the somewhat non-standard situation that in the interaction Hamiltonian one has now operators from different reservoirs occurring in a product, which implies interesting properties for the correlation functions.

5.8.2 Reservoir Equilibrium in the Polaron Picture

Before we proceed further by deriving a master equation in the displaced polaron frame, we remark that the solution from the displaced frame has to be transformed back to the original picture. A rate equation in the displaced frame implies that the full density matrix in the polaron frame is given by a product state of system and reservoir, where the phonon reservoir density matrix is given by the thermal equilibrium state $\rho'(t) = \rho'_S(t)\bar{\rho}_B^{(L)}\bar{\rho}_B^{(R)} \frac{e^{-\beta_{ph}H'_B}}{Z'_{ph}}$. The transformation back to the initial frame is given by the inverse polaron transformation

$$\rho(t) = U^\dagger \rho'(t) U = U^\dagger \rho'_S(t)\bar{\rho}_B^{(L)}\bar{\rho}_B^{(R)} UU^\dagger \frac{e^{-\beta_{ph}H'_B}}{Z'_{ph}} U$$

$$= U^\dagger \rho'_S(t) U \bar{\rho}_B^{(L)}\bar{\rho}_B^{(R)} \frac{e^{-\beta_{ph}U^\dagger H'_B U}}{Z'_{ph}}, \tag{5.150}$$

where we have used the fact that the polaron transformation (5.145) leaves the electronic reservoirs untouched. When the system density matrix does not exhibit coherences $\rho'_S(t) = P_E(t)dd^\dagger + P_F(t)d^\dagger d$, the unitary transformation will leave it untouched, such that only the reservoir part will be modified. With $H'_B = \sum_q \omega_q a_q^\dagger a_q$ we can with the inverse transformations of Eq. (5.146),

$$U^\dagger H'_B U = \sum_q \omega_q a_q^\dagger a_q + d^\dagger d \otimes \sum_q (h_q a_q + h_q^* a_q^\dagger) + \sum_q \frac{|h_q|^2}{\omega_q} d^\dagger d$$

$$= d^\dagger d \otimes \sum_q \left(\omega_q a_q^\dagger a_q + h_q a_q + h_q^* a_q^\dagger + \frac{|h_q|^2}{\omega_q}\mathbb{1}\right)$$

$$+ dd^\dagger \otimes \sum_q \omega_q a_q^\dagger a_q, \tag{5.151}$$

represent the operator in the exponential as a sum of commuting operators. Since for all operators $AB = BA = 0$ we have $e^{A+B} = e^A e^B$, we conclude that

$$e^{-\beta_{ph}U^\dagger H'_B U} = e^{-\beta_{ph}d^\dagger d \otimes \sum_q \omega_q(a_q^\dagger + h_q/\omega_q)(a_q + h_q^*/\omega_q)} e^{-\beta_{ph}dd^\dagger \otimes \sum_q \omega_q a_q^\dagger a_q}$$

$$= [1 + d^\dagger d(e^{-\beta_{ph}\sum_q \omega_q(a_q^\dagger + h_q/\omega_q)(a_q + h_q^*/\omega_q)} - 1)]$$

$$\times [1 + dd^\dagger(e^{-\beta_{ph}\sum_q \omega_q a_q^\dagger a_q} - 1)]$$

$$= d^\dagger d\, e^{-\beta_{ph}\sum_q \omega_q(a_q^\dagger + h_q/\omega_q)(a_q + h_q^*/\omega_q)} + dd^\dagger e^{-\beta_{ph}\sum_q \omega_q a_q^\dagger a_q}. \tag{5.152}$$

Comparing with the initial Hamiltonian, the phonon part of the first term in the last line is nothing but the thermal phonon state under the side constraint that the SET dot is filled. Formally, this can be seen by substituting $d^\dagger d \to 1$ in Eq. (5.144). Similarly, the other term is the thermalized phonon state when the SET dot is empty. Therefore, preparing the reservoir in a thermal state in the polaron-transformed frame implies that, in the original frame, the reservoir state is conditioned on the state of the system. Inserting the assumption that there are no coherences in the system $\rho_S'(t) = P_E(t)dd^\dagger + P_F(t)d^\dagger d$, the full density matrix in the original frame becomes

$$\rho(t) = P_E(t)dd^\dagger \bar\rho_B^{(L)} \bar\rho_B^{(R)} \otimes \frac{e^{-\beta_{ph} \sum_q \omega_q a_q^\dagger a_q}}{Z_{ph}'}$$

$$+ P_F(t)d^\dagger d\bar\rho_B^{(L)} \bar\rho_B^{(R)} \otimes \frac{e^{-\beta_{ph} \sum_q \omega_q (a_q^\dagger + h_q/\omega_q)(a_q + h_q^*/\omega_q)}}{Z_{ph}'}. \quad (5.153)$$

Therefore, when the SET dot is occupied, the phonon state is given by a displaced thermal state, whereas when the SET dot is empty, it is just given by the thermal state corresponding to the original phonon Hamiltonian. The phonon dynamics thereby follows the system state immediately, which goes beyond the conventional Born approximation.

5.8.3 Polaron Rate Equation for Discrete Phonon Modes

In the transformed frame, we now proceed to derive a rate equation for the SET dot populations. Identifying the bath coupling operators in the interaction Hamiltonian (5.149) as

$$B_{1\alpha} = \sum_k t_{k\alpha} c_{k\alpha}^\dagger e^{-A}, \qquad B_{2\alpha} = \sum_k t_{k\alpha}^* c_{k\alpha} e^{+A}, \quad (5.154)$$

it becomes quite obvious that the reservoir correlation functions will now simultaneously contain contributions from the electronic and phonon reservoirs. In the time domain, these enter in simple product form:

$$C_{12}^\alpha(\tau) = \langle B_{1\alpha}(\tau) B_{2\alpha} \rangle = C_{12,el}^\alpha(\tau) C_{ph}(\tau),$$

$$C_{21}^\alpha(\tau) = \langle B_{2\alpha}(\tau) B_{1\alpha} \rangle = C_{21,el}^\alpha(\tau) C_{ph}(\tau), \quad (5.155)$$

where the phonon contribution is given by

$$C_{ph}(\tau) = \langle e^{-A(\tau)} e^{+A} \rangle = \langle e^{+A(\tau)} e^{-A} \rangle \quad (5.156)$$

with the phonon operator in the interaction picture

$$A(\tau) = \sum_q \left(\frac{h_q^*}{\omega_q} a_q^\dagger e^{+i\omega_q \tau} - \frac{h_q}{\omega_q} a_q e^{-i\omega_q \tau} \right). \quad (5.157)$$

The electronic contributions are just the conventional ones known from the SET,

$$C_{12,\text{el}}(\tau) = \sum_k |t_{k\alpha}|^2 f_\alpha(\varepsilon_{k\alpha}) e^{+i\varepsilon_{k\alpha}\tau} = \frac{1}{2\pi} \int \Gamma_\alpha(-\omega) f_\alpha(-\omega) e^{-i\omega\tau} \, d\omega,$$

$$C_{21,\text{el}}(\tau) = \sum_k |t_{k\alpha}|^2 \big[1 - f_\alpha(\varepsilon_{k\alpha})\big] e^{-i\varepsilon_{k\alpha}\tau} = \frac{1}{2\pi} \int \Gamma_\alpha(\omega)\big[1 - f_\alpha(\omega)\big] e^{-i\omega\tau} \, d\omega.$$

$$(5.158)$$

Compare also Eqs. (5.3), (5.4), and (5.5).

To calculate the phonon contribution to the correlation function, we can exploit that

$$\big[A(\tau), A\big] = 2i \sum_q \frac{|h_q|^2}{\omega_q^2} \sin(\omega_q\tau) \tag{5.159}$$

is just a number, which implies, using the Baker–Campbell–Hausdorff relation,

$$e^{-A(\tau)}e^{+A} = e^{A - A(\tau) - 1/2[A(\tau),A]}$$

$$= e^{\sum_q \left(\frac{h_q^*}{\omega_q} a_q^\dagger (1 - e^{+i\omega_q\tau}) - \frac{h_q}{\omega_q} a_q (1 - e^{-i\omega_q\tau})\right)} e^{-i\sum_q \frac{|h_q|^2}{\omega_q^2} \sin(\omega_q\tau)}. \tag{5.160}$$

For a thermal reservoir, the phonon correlation function can be written as a product of single-mode correlation functions $C_{\text{ph}}(\tau) = \prod_{q=1}^Q C_{\text{ph}}^q(\tau)$, where the single-mode contributions read

$$C_{\text{ph}}^q(\tau) = \left\langle e^{\frac{h_q^*}{\omega_q} a_q^\dagger (1 - e^{+i\omega_q\tau}) - \frac{h_q}{\omega_q} a_q (1 - e^{-i\omega_q\tau})} e^{-i\frac{|h_q|^2}{\omega_q^2} \sin(\omega_q\tau)} \right\rangle$$

$$= \left\langle e^{\frac{h_q^*}{\omega_q} a_q^\dagger (1 - e^{+i\omega_q\tau})} e^{-\frac{h_q}{\omega_q} a_q (1 - e^{-i\omega_q\tau})} \right\rangle e^{-\frac{|h_q|^2}{\omega_q^2} (1 - e^{-i\omega_q\tau})}. \tag{5.161}$$

By expanding the exponentials, we can evaluate the expectation value for thermal states, where the probability of having n quanta in the mode q is given by $P_n = (1 - e^{-\beta_{\text{ph}}\omega_q})e^{-n\beta_{\text{ph}}\omega_q}$ as

$$\left\langle e^{\alpha_q^* a_q^\dagger} e^{-\alpha_q a_q} \right\rangle = \sum_{n,m=0}^\infty \frac{(\alpha_q^*)^n}{n!} \frac{(-\alpha_q)^m}{m!} \sum_{\ell=0}^\infty P_\ell \langle \ell | (a_q^\dagger)^n (a_q)^m | \ell \rangle$$

$$= \sum_{n=0}^\infty \frac{(-1)^n |\alpha_q|^{2n}}{(n!)^2} \sum_{\ell=0}^\infty P_\ell \langle \ell | (a_q^\dagger)^n (a_q)^n | \ell \rangle$$

$$= \sum_{\ell=0}^{\infty} P_\ell \sum_{n=0}^{\ell} \frac{(-1)^n |\alpha_q|^{2n}}{(n!)^2} \frac{\ell!}{(\ell-n)!}$$

$$= \sum_{\ell=0}^{\infty} P_\ell \mathscr{L}_\ell (|\alpha_q|^2) = e^{-|\alpha_q|^2 n_B^q} \tag{5.162}$$

with the Bose distribution $n_B^q = [e^{\beta_{\mathrm{ph}}\omega_q} - 1]^{-1}$ and Legendre polynomials, defined by the Rodrigues formula [25]

$$\mathscr{L}_n(x) = \frac{1}{2^n n!} \frac{d^n}{dx^n} [x^2 - 1]^n. \tag{5.163}$$

The single-mode contributions thus become with $\alpha_q = \frac{h_q}{\omega_q}(1 - e^{-i\omega_q \tau})$

$$C_{\mathrm{ph}}^q = \exp\left\{ \frac{|h_q|^2}{\omega_q^2} \left[e^{-i\omega_q \tau}(1 + n_B^q) + e^{+i\omega_q \tau} n_B^q - (1 + 2n_B^q) \right] \right\}, \tag{5.164}$$

such that finally, we obtain for the phonon correlation function

$$C_{\mathrm{ph}}(\tau) = \exp\left\{ \sum_q \frac{|h_q|^2}{\omega_q^2} \left[e^{-i\omega_q \tau}(1 + n_B^q) + e^{+i\omega_q \tau} n_B^q - (1 + 2n_B^q) \right] \right\}. \tag{5.165}$$

The fact that the transformation $h_q \to -h_q$ leaves this result invariant implies that the phonon contribution is always the same in Eq. (5.155). We note that the phonon correlation function obeys the KMS condition (2.51).

Exercise 5.15 (KMS condition) Show that the phonon correlation function (5.165) obeys the KMS condition $C(\tau) = C(-\tau - i\beta_{\mathrm{ph}})$.

The observation that in the phonon correlation function (5.164) the terms proportional to $(1 + n_B^q)$ correspond to the emission of a phonon into the phonon reservoir and terms proportional to n_B^q alone are responsible for the absorption of a phonon from the reservoir enables one to derive the full phonon counting statistics from the model. Formally expanding the single-mode correlation function into multiple emission (m') and absorption (m) events,

$$C_{\mathrm{ph}}^q(\tau) = e^{-\frac{|h_q|^2}{\omega_q^2}(1+2n_B^q)}$$

$$\times \sum_{m,m'=0}^{\infty} \left(\frac{|h_q|^2}{\omega_q^2} \right)^{m+m'} \frac{(n_B^q)^m (1+n_B^q)^{m'}}{m! m'!} e^{+i(m-m')\omega_q \tau}, \tag{5.166}$$

one can show that by introducing the net number of phonon absorptions by the phonon bath $n = m' - m$, the correlation function can be represented as

$$C_{\text{ph}}^q(\tau) = \sum_{n=-\infty}^{+\infty} e^{-in\omega_q \tau} e^{-\frac{|h_q|^2}{\omega_q^2}(1+2n_B^q)} \left(\frac{1+n_B^q}{n_B^q}\right)^{\frac{n}{2}}$$

$$\times \mathscr{I}_n\left(2\frac{|h_q|^2}{\omega_q^2}\sqrt{n_B^q(1+n_B^q)}\right), \tag{5.167}$$

where $\mathscr{I}_n(x)$ denotes the modified Bessel function of the first kind [25]—defined by the solution of the differential equation $z^2 \mathscr{I}_n''(z) + z\mathscr{I}_n'(z) - (z^2+n^2)\mathscr{I}_n(z) = 0$. Introducing for multiple modes the notation $\boldsymbol{n} = (n_1, \ldots, n_Q)$, $\boldsymbol{\omega} = (\omega_1, \ldots, \omega_Q)$, we therefore have for the full multi-mode phonon correlation function the representation

$$C_{\text{ph}}(\tau) = \sum_{\boldsymbol{n}} e^{-i\boldsymbol{n}\cdot\boldsymbol{\omega}\tau} \prod_{q=1}^{Q} \left[e^{-\frac{|h_q|^2}{\omega_q^2}(1+2n_B^q)} \left(\frac{1+n_B^q}{n_B^q}\right)^{\frac{n_q}{2}} \mathscr{I}_{n_q}\left(2\frac{|h_q|^2}{\omega_q^2}\sqrt{n_B^q(1+n_B^q)}\right) \right]$$

$$= \sum_{\boldsymbol{n}} e^{-i\boldsymbol{n}\cdot\boldsymbol{\omega}\tau} C_{\text{ph}}^{\boldsymbol{n}}, \tag{5.168}$$

where the simple exponential prefactor enables us to calculate the Fourier transform of the full correlation function. In particular, if only a single phonon mode is present, this enables a simple calculation of the Fourier transform of the complete electron–phonon correlation function:

$$\gamma_{12}^\alpha(\omega) = \sum_{\boldsymbol{n}_\alpha} \gamma_{12,\text{el}}^\alpha(\omega - \boldsymbol{n}_\alpha \cdot \boldsymbol{\omega})C_{\text{ph}}^{\boldsymbol{n}_\alpha} = \sum_{\boldsymbol{n}_\alpha} \gamma_{12,\boldsymbol{n}_\alpha}^\alpha(\omega),$$

$$\gamma_{21}^\alpha(\omega) = \sum_{\boldsymbol{n}_\alpha} \gamma_{21,\text{el}}^\alpha(\omega - \boldsymbol{n}_\alpha \cdot \boldsymbol{\omega})C_{\text{ph}}^{\boldsymbol{n}_\alpha} = \sum_{\boldsymbol{n}_\alpha} \gamma_{21,\boldsymbol{n}_\alpha}^\alpha(\omega). \tag{5.169}$$

Here, the terms $\gamma_{12,\boldsymbol{n}_\alpha}^\alpha$ are interpreted as the emission of $\boldsymbol{n}_\alpha$ phonons into the phonon reservoir while an electron jumps from lead α onto the SET dot, whereas $\gamma_{21,\boldsymbol{n}_\alpha}^\alpha$ accounts for the emission of $\boldsymbol{n}_\alpha$ when an electron is emitted to lead α. Now, the bosonic KMS relation

$$C_{\text{ph}}^{-\boldsymbol{n}_\alpha} = e^{-\beta_{\text{ph}}\boldsymbol{n}_\alpha \cdot \boldsymbol{\omega}} C_{\text{ph}}^{+\boldsymbol{n}_\alpha} \tag{5.170}$$

together with properties of the Fermi functions imply a KMS-type relation for the full correlation function

$$\gamma_{12,+\boldsymbol{n}_\alpha}^\alpha(-\omega) = e^{-\beta_\alpha(\omega - \mu_\alpha + \boldsymbol{n}_\alpha \cdot \boldsymbol{\omega})} e^{+\beta_{\text{ph}}\boldsymbol{n}_\alpha \cdot \boldsymbol{\omega}} \gamma_{21,-\boldsymbol{n}_\alpha}^\alpha(+\omega), \tag{5.171}$$

which now involves both the electronic and phononic temperatures.

Exercise 5.16 (KMS condition) Show the validity of relation (5.171).

However, we note that when these temperatures are equal, the usual local detailed balance relations are reproduced. Deriving a rate equation (2.47) for the dot occupation is now straightforward. The probabilities for finding the dot empty or filled are governed by the rate matrix

$$\mathscr{L} = \sum_{\alpha \in \{L,R\}} \sum_{n_\alpha} \begin{pmatrix} -\gamma^\alpha_{12,n_\alpha}(-\varepsilon') & +\gamma^\alpha_{21,-n_\alpha}(+\varepsilon') \\ +\gamma^\alpha_{12,n_\alpha}(-\varepsilon') & -\gamma^\alpha_{21,-n_\alpha}(+\varepsilon') \end{pmatrix},$$

where $\gamma^\alpha_{12,n_\alpha}(-\varepsilon')$ denotes the rate for an electron jumping onto the SET dot from lead α while simultaneously emitting n_α phonons of the various modes into the phonon reservoir. Correspondingly, $\gamma^\alpha_{21,-n_\alpha}(+\varepsilon')$ denotes the rate for the inverse process. Having identified the rates for the various involved processes, we can proceed by introducing counting fields. For a three-terminal system with the phononic junction only allowing for energy exchange and with conservation laws on the total energy and particle number, we can expect three counting fields to be sufficient for tracking the full entropy production. These can be, e.g., the matter transfer from left to right and the energy emitted to the phonon bath counted separately for left and right electronic jumps, such that we have the counting-field-dependent version

$$\mathscr{L}(\chi,\xi_L,\xi_R)$$
$$= \begin{pmatrix} -\gamma^L_{12,n_L}(-\varepsilon') & +\gamma^L_{21,-n_L}(+\varepsilon')e^{-\mathrm{i}n_L\cdot\Omega\xi_L} \\ +\gamma^L_{12,n_L}(-\varepsilon')e^{+\mathrm{i}n_L\cdot\Omega\xi_L} & -\gamma^L_{21,-n_L}(+\varepsilon') \end{pmatrix}$$
$$+ \begin{pmatrix} -\gamma^R_{12,n_R}(-\varepsilon') & +\gamma^R_{21,-n_R}(+\varepsilon')e^{+\mathrm{i}\chi}e^{-\mathrm{i}n_R\cdot\Omega\xi_R} \\ +\gamma^R_{12,n_R}(-\varepsilon')e^{-\mathrm{i}\chi}e^{+\mathrm{i}n_R\cdot\Omega\xi_R} & -\gamma^L_{21,-n_R}(+\varepsilon') \end{pmatrix},$$

which enables one to reconstruct all energy and matter currents and thus the full entropy flow.

Here, we will first investigate the impact of the phonon presence on the electronic matter current. If one is only interested in the electronic current, we may set $\xi_L = \xi_R = 0$. The transition rates in the above Liouvillian become particularly simple in the case of a single phonon mode:

$$\gamma^\alpha_{12,+n}(-\varepsilon') = \Gamma_\alpha(\varepsilon'+n\Omega)f_\alpha(\varepsilon'+n\Omega)e^{-\Lambda(1+2n_B)}\left(\frac{1+n_B}{n_B}\right)^{\frac{n}{2}}$$
$$\times \mathscr{J}_n\left(2\Lambda\sqrt{n_B(1+n_B)}\right),$$

$$\gamma^\alpha_{21,-n}(+\varepsilon') = \Gamma_\alpha(\varepsilon'+n\Omega)\left[1-f_\alpha(\varepsilon'+n\Omega)\right]e^{-\Lambda(1+2n_B)}\left(\frac{n_B}{1+n_B}\right)^{\frac{n}{2}}$$
$$\times \mathscr{J}_n\left(2\Lambda\sqrt{n_B(1+n_B)}\right),$$

(5.172)

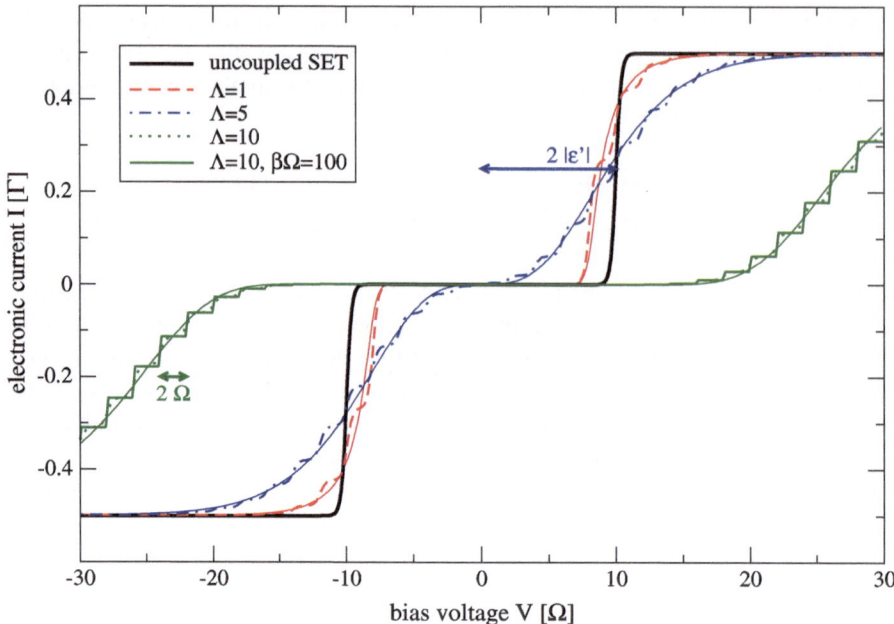

Fig. 5.20 Electronic matter current versus bias voltage applied to the SET for vanishing (*bold black*) and increasing (*dashed red, dash–dotted blue,* and *dotted green,* respectively) coupling strengths $\Lambda = |h|^2/\Omega^2 = J_0$ to a single phonon mode of frequency Ω (*bold curves*) or to a continuum of phonon modes distributed according to an ohmic model (*thin solid curves* in *background*). The Franck–Condon blockade can be understood within this model in terms of a renormalization of the effective dot level $\varepsilon' = \varepsilon - \Lambda\Omega$, which, when $\Lambda\Omega \gg \varepsilon$, will lead to current suppression. Furthermore, the steps in the electronic current observed for sufficiently low temperatures (*solid green*) admit the transport spectroscopy of the phonon frequency Ω. In the multi-mode case (*thin solid curves,* for $\omega_c = \Omega$ and $J_0 = \Lambda$), current suppression due to the level renormalization is also observed, but the steps in the current are no longer visible. Other parameters: $\Gamma_L = \Gamma_R = \Gamma$, $\beta_L = \beta_R = \beta_{\mathrm{ph}} = \beta$, $\beta\Omega = 10$ (except the *thin green curve*), $\varepsilon = 5\Omega$, $J_0 = \Lambda$, $\omega_c = \Omega$

where $\Lambda = \frac{|h|^2}{\Omega_q^2}$ denotes the dimensionless coupling strength to the single phonon mode which is occupied according to $n_B = [e^{\beta_{\mathrm{ph}}\Omega} - 1]^{-1}$. The resulting electronic matter current is depicted in Fig. 5.20. Surprisingly, the simple 2×2 rate matrix predicts many signatures in the electronic current. For example, in the electronic matter current one can read off the renormalized dot level at sufficiently low electronic temperatures. In addition however, low temperatures also allow one to determine the phonon frequency from the width of the multiple plateaus.

5.8.4 Polaron Rate Equation for Continuum Phonon Modes

It is also possible to obtain a master equation representation for a continuum of phonon modes. Here, we directly represent the phonon correlation function (5.165) by

$$C_{\mathrm{ph}}(\tau) = \exp\left\{ \int_0^\infty d\omega \, \frac{J(\omega)}{\omega^2} \left[e^{-i\omega\tau}\left(1 + n_B(\omega)\right) + e^{+i\omega\tau} n_B(\omega) - \left(1 + 2n_B(\omega)\right) \right] \right\},$$

(5.173)

where we have introduced the spectral density $J(\omega) = \sum_q |h_q|^2 \delta(\omega - \omega_q)$. When we choose the common ohmic parametrization $J(\omega) = J_0 \omega e^{-\omega/\omega_c}$ with dimensionless coupling strength J_0 and cutoff frequency ω_c, the integral can be solved exactly. Writing the Bose–Einstein distributions as a geometric series and again summing all separate integral contributions, we finally obtain for the phonon correlation function

$$C_{\mathrm{ph}}(\tau) = \left[\frac{\Gamma(\frac{1+\beta_{\mathrm{ph}}\omega_c + i\tau\omega_c}{\beta_{\mathrm{ph}}\omega_c}) \Gamma(\frac{1+\beta_{\mathrm{ph}}\omega_c - i\tau\omega_c}{\beta_{\mathrm{ph}}\omega_c})}{\Gamma^2(\frac{1+\beta_{\mathrm{ph}}\omega_c}{\beta_{\mathrm{ph}}\omega_c})(1 + i\tau\omega_c)} \right]^{J_0},$$

(5.174)

where $\Gamma(x) = \int_0^\infty t^{x-1} e^{-t} dt$ denotes the Γ-function. We note from Eq. (5.173) that for particular parametrizations of the spectral coupling density one can expect that for large times the phonon correlation functions may remain finite with $\lim_{t\to\infty} C_{\mathrm{ph}}(\tau) \neq 0$. However, the total correlation function is given by a product of electronic functions (which decay) and phonon correlation functions. Its Fourier transform (which enters the rates) can be calculated numerically from a convolution integral:

$$\gamma_{12}^\alpha(-\varepsilon') = \frac{1}{2\pi} \int d\omega \, \Gamma_\alpha(-\omega) f_\alpha(-\omega) \gamma_{\mathrm{ph}}(-\varepsilon' - \omega),$$

$$\gamma_{21}^\alpha(+\varepsilon') = \frac{1}{2\pi} \int d\omega \, \Gamma_\alpha(+\omega)\left[1 - f_\alpha(+\omega)\right] \gamma_{\mathrm{ph}}(+\varepsilon' - \omega),$$

(5.175)

and enters in this case a rate matrix of the form

$$\mathcal{L}(\chi) = \begin{pmatrix} -\gamma_{12}^L(-\varepsilon') & +\gamma_{21}^L(+\varepsilon') \\ +\gamma_{12}^L(-\varepsilon') & -\gamma_{21}^L(+\varepsilon') \end{pmatrix}$$
$$+ \begin{pmatrix} -\gamma_{12}^R(-\varepsilon') & +\gamma_{21}^R(+\varepsilon')e^{+i\chi} \\ +\gamma_{12}^R(-\varepsilon')e^{-i\chi} & -\gamma_{21}^R(+\varepsilon') \end{pmatrix},$$

(5.176)

from which the electronic matter current can be directly deduced. With the choices $J_0 = \frac{|h|^2}{\Omega^2}$ and $\omega_c = \Omega$ the electronic current for high temperatures is quite similar, as if one would have only a single phonon mode. The crucial difference however is that at low temperatures, the phonon plateaus are no longer visible—compare the thin solid versus the bold curves in Fig. 5.20. Since for the continuum model many different phonon frequencies contribute, this is expected. Interestingly however, the current suppression due to the presence of the phonons (Franck–Condon blockade) is also visible for a continuum of phonon modes.

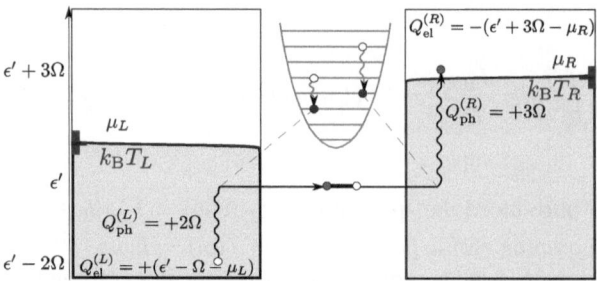

Fig. 5.21 Sketch of the energetics of the problem for a single phonon mode, slightly adapted from Ref. [3]. For sufficiently low electronic temperatures, the dot level must be between μ_L and μ_R to allow for transport, such that an electronic transfer from left to right would be extremely unlikely for the depicted situation. With phonons at sufficiently large temperature however, it is possible to realize trajectories where the missing energy is supplied by the phonon bath. The indicated heat transfers from reservoirs into the system allow for a complete reconstruction of the entropy flows, even for single trajectories

5.8.5 Thermodynamic Interpretation

The present rate equation does not directly fit the scheme in Sect. 4.3, since the contribution of the three reservoirs to the rates is not additive. Nevertheless, an interpretation in terms of stochastic thermodynamics is possible.

The strong modification of the electronic current is due to the fact that the phonons allow for processes that would normally be forbidden; see Fig. 5.21. In the trajectory in the figure, first an electron jumps in from the left lead to the initially empty SET while absorbing two phonons. The change of the system energy by $\Delta E = +\varepsilon' = \Delta E_L + \Delta E_{\mathrm{ph}}$ is supplied by both the left lead $\Delta E_L = \varepsilon' - 2\Omega$ and the phonon bath $\Delta E_{\mathrm{ph}} = +2\Omega$. In the second step, the electron leaves the dot towards the right lead while again absorbing three phonons. Again, the change of the system energy by $-\varepsilon'$ is supplied by the right lead $\Delta E_R = -(\varepsilon' + 3\Omega)$ and the phonon bath $\Delta E_{\mathrm{ph}} = +3\Omega$. These energy and matter transfers can be used to construct the total heat exchanged between the reservoirs and thereby also the total entropy production in the steady state.

To relate the thermodynamic interpretation more to the modified local detailed balance relation, let us now for simplicity restrict ourselves to the case of a single phonon mode (the generalization to multiple modes is also possible). Formally, the rates corresponding to emission or absorption of different phonon numbers enter additively in Eq. (5.172). This enables one to see the phonon reservoir as a whole collection of infinitely many virtual phonon reservoirs that admit only the emission or absorption of a certain number of phonons with the same frequency each time an electron is transferred across the SET junctions. This view enables one to adopt the definition of the entropy flow Definition 4.3, where the index ν labeling the reservoir may now assume infinitely many values $\nu = (\alpha, n)$, where $\alpha \in \{L, R\}$ denotes the junction across which an electron is transferred and n denotes the virtual phonon reservoir from or to which only n phonons may be absorbed or emitted. Recalling that $\mathscr{L}_{EF}^{(\alpha,n)}$ denotes the rate for an electron to leave the dot towards lead α while absorbing n phonons from the reservoir and $\mathscr{L}_{FE}^{(\alpha,n)}$ the rate of the inverse process, i.e., for an electron to enter the dot from lead α while emitting n phonons into the reservoir, the local detailed balance relation becomes, with the rates in Eq. (5.172),

$$\ln\left(\frac{\mathscr{L}_{FE}^{(\alpha,n)}}{\mathscr{L}_{EF}^{(\alpha,n)}}\right) = \ln\left(\frac{\gamma_{12,+n}(-\varepsilon')}{\gamma_{21,-n}(+\varepsilon')}\right) = \ln\left[\frac{f_\alpha(\varepsilon'+n\Omega)}{1-f_\alpha(\varepsilon'+n\Omega)}\left(\frac{1+n_B}{n_B}\right)^n\right]$$

$$= \ln\left[e^{-\beta_\alpha(\varepsilon'+n\Omega-\mu_\alpha)}e^{+n\beta_{\mathrm{ph}}\Omega}\right]$$

$$= -\beta_\alpha\left(\varepsilon'+n\Omega-\mu_\alpha\right)+\beta_{\mathrm{ph}}n\Omega, \tag{5.177}$$

such that the entropy flow from the virtual reservoir $\nu=(\alpha,n)$ becomes

$$\dot{S}_{\mathrm{e}}^{(\alpha,n)} = \mathscr{L}_{EF}^{(\alpha,n)}\bar{P}_F\ln\left(\frac{\mathscr{L}_{FE}^{(\alpha,n)}}{\mathscr{L}_{EF}^{(\alpha,n)}}\right)+\mathscr{L}_{FE}^{(\alpha,n)}\bar{P}_E\ln\left(\frac{\mathscr{L}_{EF}^{(\alpha,n)}}{\mathscr{L}_{FE}^{(\alpha,n)}}\right)$$

$$= \left[\mathscr{L}_{EF}^{(\alpha,n)}\bar{P}_F-\mathscr{L}_{FE}^{(\alpha,n)}\bar{P}_E\right]\ln\left(\frac{\mathscr{L}_{FE}^{(\alpha,n)}}{\mathscr{L}_{EF}^{(\alpha,n)}}\right)$$

$$= \beta_\alpha\left(I_E^{(\alpha,n)}-\mu_\alpha I_M^{(\alpha,n)}\right)+\beta_{\mathrm{ph}}I_E^{(n,\alpha,\mathrm{ph})} = \dot{S}_{\mathrm{e,el}}^{(\alpha,n)}+\dot{S}_{\mathrm{e,ph}}^{(\alpha,n)}, \tag{5.178}$$

which is additive in electronic and phononic contributions. Here, we have introduced the energy flows corresponding to the emission or absorption of n phonons. The total energy flows are given by

$$I_E^\alpha = \sum_n I_E^{(\alpha,n)} = \sum_n\left[\gamma_{12,+n}(-\varepsilon')\bar{P}_E-\gamma_{21,-n}(+\varepsilon')\bar{P}_F\right](\varepsilon'+n\Omega),$$

$$I_E^{\mathrm{ph}} = \sum_n\left[I_E^{(n,L,\mathrm{ph})}+I_E^{(n,R,\mathrm{ph})}\right] \tag{5.179}$$

$$= \sum_n\sum_\alpha\left[\gamma_{21,-n}(+\varepsilon')\bar{P}_F-\gamma_{12,+n}(-\varepsilon')\bar{P}_R\right]n\Omega,$$

whereas the total electronic matter current from lead α is given by

$$I_M^\alpha = \sum_n I_E^{(\alpha,n)} = \sum_n\left[\gamma_{12,+n}(-\varepsilon')\bar{P}_E-\gamma_{21,-n}(+\varepsilon')\bar{P}_F\right]. \tag{5.180}$$

Similarly, the total entropy flow from the electronic leads is obtained by summing over all different n, and the total entropy flow from the phonon reservoirs is obtained by summing over the contributions from different n and different α:

$$\dot{S}_{\mathrm{e}}^{(\alpha)} = \sum_n\dot{S}_{\mathrm{e,el}}^{(\alpha,n)},$$

$$\dot{S}_{\mathrm{e}}^{\mathrm{ph}} = \sum_n\left(\dot{S}_{\mathrm{e,ph}}^{(L,n)}+\dot{S}_{\mathrm{e,ph}}^{(R,n)}\right). \tag{5.181}$$

Altogether, the system obeys the laws of thermodynamics, which results in an overall positive entropy production. Consequently, we just note here that it is possible to verify a fluctuation theorem for entropy production, i.e., for $P_{n,e^L_{\mathrm{ph}},e^R_{\mathrm{ph}}}(t)$ denoting the probability for trajectories with n electrons having traversed the SET from left to right and having emitted energy $e^L_{\mathrm{ph}} = n_L \cdot \omega$ to the phonon reservoir during electronic jumps over the left and energy $e^R_{\mathrm{ph}} = n_R \cdot \omega$ during jumps over the right barrier. In detail, it reads [3]

$$\lim_{t\to\infty} \frac{P_{+n,+e^L_{\mathrm{ph}},+e^R_{\mathrm{ph}}}(t)}{P_{-n,-e^L_{\mathrm{ph}},-e^R_{\mathrm{ph}}}(t)}$$

$$= e^{[(\beta_R-\beta_L)\varepsilon'+(\beta_L\mu_L-\beta_R\mu_R)]n+(\beta_{\mathrm{ph}}-\beta_L)e^L_{\mathrm{ph}}+(\beta_{\mathrm{ph}}-\beta_R)e^R_{\mathrm{ph}}}, \tag{5.182}$$

and it is straightforward to see that it reduces to the conventional fluctuation theorem when all temperatures are equal.

Disregarding the phonon counting statistics, we note that the system also obeys a fluctuation theorem involving the electronic transfer statistics only,

$$\lim_{t\to\infty} \frac{P_{+n}(t)}{P_{-n}(t)} = e^{n\mathscr{A}_{\mathrm{eff}}}, \tag{5.183}$$

where the effective affinity $\mathscr{A}_{\mathrm{eff}}$ is however not related to the entropy production.

References

1. H. Haug, A.-P. Jauho, *Quantum Kinetics in Transport and Optics of Semiconductors* (Springer, Berlin, 2008)
2. M. Esposito, K. Lindenberg, C.V. den Broeck, Thermoelectric efficiency at maximum power in a quantum dot. Europhys. Lett. **85**, 60010 (2009)
3. G. Schaller, T. Krause, T. Brandes, M. Esposito, Single-electron transistor strongly coupled to vibrations: counting statistics and fluctuation theorem. New J. Phys. **15**, 033032 (2013)
4. T. Krause, G. Schaller, T. Brandes, Incomplete current fluctuation theorems for a four-terminal model. Phys. Rev. B **84**, 195113 (2011)
5. G. Schaller, G. Kießlich, T. Brandes, Low-dimensional detector model for full counting statistics: trajectories, back action, and fidelity. Phys. Rev. B **82**, 041303 (2010)
6. J. Mehl, B. Lander, C. Bechinger, V. Blickle, U. Seifert, Role of hidden slow degrees of freedom in the fluctuation theorem. Phys. Rev. Lett. **108**, 220601 (2012)
7. P. Strasberg, G. Schaller, T. Brandes, M. Esposito, Thermodynamics of a physical model implementing a maxwell demon. Phys. Rev. Lett. **110**, 040601 (2013)
8. G.B. Cuetara, M. Esposito, G. Schaller, P. Gaspard, Effective fluctuation theorems for electron transport in a double quantum dot coupled to a quantum point contact. Phys. Rev. B **88**, 115134 (2013)
9. D.S. Golubev, Y. Utsumi, M. Marthaler, G. Schön, Fluctuation theorem for a double quantum dot coupled to a point-contact electrometer. Phys. Rev. B **84**, 075323 (2011)
10. Y. Utsumi, D.S. Golubev, M. Marthaler, K. Saito, T. Fujisawa, G. Schön, Bidirectional single-electron counting and the fluctuation theorem. Phys. Rev. B **81**, 125331 (2010)

11. G. Bulnes Cuetara, M. Esposito, P. Gaspard, Fluctuation theorems for capacitively coupled electronic currents. Phys. Rev. B **84**, 165114 (2011)
12. M. Esposito, Stochastic thermodynamics under coarse graining. Phys. Rev. E **85**, 041125 (2012)
13. S. Gustavsson, R. Leturcq, B. Simovic, R. Schleser, T. Ihn, P. Studerus, K. Ensslin, D.C. Driscoll, A.C. Gossard, Counting statistics of single electron transport in a quantum dot. Phys. Rev. Lett. **96**, 076605 (2006)
14. T. Fujisawa, T. Hayashi, R. Tomita, Y. Hirayama, Bidirectional counting of single electrons. Science **312**, 1634 (2006)
15. R. Sánchez, R. Lopez, D. Sanchez, M. Büttiker, Mesoscopic Coulomb drag, broken detailed balance, and fluctuation relations. Phys. Rev. Lett. **104**, 076801 (2010)
16. R. Hussein, S. Kohler, Coherent quantum ratchets driven by tunnel oscillations: fluctuations and correlations. Phys. Rev. B **86**, 115452 (2012)
17. C. Flindt, C. Fricke, F. Hohls, T. Novotny, K. Netocny, T. Brandes, R.J. Haug, Universal oscillations in counting statistics. Proc. Natl. Acad. Sci. USA **106**, 10116 (2009)
18. G. Kießlich, E. Schöll, T. Brandes, F. Hohls, R.J. Haug, Noise enhancement due to quantum coherence in coupled quantum dots. Phys. Rev. Lett. **99**, 206602 (2007)
19. B. Rutten, M. Esposito, B. Cleuren, Reaching optimal efficiencies using nanosized photoelectric devices. Phys. Rev. B **80**, 235122 (2009)
20. C.V. den Broeck, Thermodynamic efficiency at maximum power. Phys. Rev. Lett. **95**, 190602 (2005)
21. F.L. Curzon, B. Ahlborn, Efficiency of a Carnot engine at maximum power output. Am. J. Phys. **43**, 22 (1975)
22. J. Koch, F. von Oppen, Franck–Condon blockade and giant Fano factors in transport through single molecules. Phys. Rev. Lett. **94**, 206804 (2005)
23. G.D. Mahan, *Many-Particle Physics* (Springer, Amsterdam, 2000)
24. T. Brandes, Coherent and collective quantum optical effects in mesoscopic systems. Phys. Rep. **408**, 315 (2005)
25. M. Abramowitz, I.A. Stegun (eds.), Handbook of Mathematical Functions. National Bureau of Standards, 1970

Chapter 6
Piecewise Constant Control

Abstract This chapter provides the theoretical framework for external control applied to a master equation formalism, when the control parameters are changed in a piecewise constant fashion. Both open-loop and closed-loop (feedback) control are discussed, and it is shown how the Wiseman–Milburn control can be expressed using Kraus maps.

The possibility of altering the quantum dynamics of a system by applying some time-dependent control is intriguing [1, 2]. Applications include, e.g., the inhibition or reduction of decoherence [3] or the construction of single-electron pumps as reliable charge emitters [4]. Unfortunately, only few time dependencies lead to analytically solvable models [5]. This is for example possible in the adiabatic limit [6], where the time dependencies are so slow that the adiabatic theorem applies, and the equations from the static configurations are mostly just equipped with time-dependent parameters [7]. Another interesting class of analytically treatable time dependencies is that of the periodic limit, where the external control function varies periodically in time, such that Floquet theory applies [8]. However, many more time dependencies are conceivable and realizable.

Here, we will explore a rather simple approach to time-dependent control: piecewise constant evolution of control parameters [9]. Such control schemes do not require any special mathematical formalism: having a system subject to a constant parameter set evolved to time t_s, one can—when a control parameter is switched instantaneously from one value to another—take the solution at time t_s as the new initial state and evolve it further until the next switching event occurs. This is the most trivial extension of theories with time-independent parameters to those with time-dependent ones, and the requirement of instantaneous switching is quite restrictive. Though an abrupt change of a parameter cannot be realized exactly in an experiment, it may be well approximated when the change is much faster than any other intrinsic time scale of the considered system. As a great advantage, these schemes may yield quite simple models that allow interpretation of the results in a physically meaningful way. The insight gained from these simple models may serve as a guiding intuition when advancing to more complex controls.

Also with parameters that change in a piecewise constant fashion in time, one should generally distinguish between two kinds of control. Open-loop control just

G. Schaller, *Open Quantum Systems Far from Equilibrium*, Lecture Notes in Physics 881,
DOI 10.1007/978-3-319-03877-3_6,
© Springer International Publishing Switzerland 2014

follows a predefined protocol. Here, the time dependence of the parameters is independent of the actual system state, and thus no knowledge of the actual system state is necessary to perform the scheme. The system is then also incapable of responding appropriately to unforeseen events. In contrast, in closed-loop (feedback) control the control actions depend on the state of the system prior to the control action. This requires measurement of the state and signal processing before the actual control action is taken—which is then of course dependent on the system state. When the latter is applied to quantum systems, we therefore also have to incorporate the measurement postulate into the evolution.

This chapter provides a basic framework for piecewise constant control schemes in a nutshell; however, for some applications the reader is referred to Chap. 7.

6.1 Piecewise Constant Open-Loop Control

For a system of ordinary differential equations with constant coefficients such as the Lindblad master equation $\dot{\rho} = \mathscr{L}_p \rho$, where the subscript p denotes the parameter dependence of the rate matrix $\mathscr{L}$, the propagator for a time interval Δt is readily given by

$$\mathscr{P}_p(\Delta t) = e^{\mathscr{L}_p \Delta t}. \tag{6.1}$$

If we now consider regular switchings at time intervals Δt, the used parameters for each interval define a control protocol $\{p_1, p_2, \ldots, p_N\}$. Altogether, the action of the combined control protocol can simply be expressed by the product of the separate control propagators. The total propagator becomes

$$\mathscr{P}_{\text{protocol}}(N \Delta t) = \mathscr{P}_{p_N}(\Delta t) \mathscr{P}_{p_{N-1}}(\Delta t) \cdots \mathscr{P}_{p_2}(\Delta t) \mathscr{P}_{p_1}(\Delta t), \tag{6.2}$$

which evolves the system state from t to $t + N \Delta t$. The propagator for a whole protocol may even be amenable to analytic investigations. For example, when the protocol only involves periodic switching between just two parameter sets (e.g., turnstyle protocols), one may simplify the total propagator using the Baker–Campbell–Hausdorff (BCH) formula. Alternatively, when the switching is performed very fast, $|\mathscr{L}_p \Delta t| \ll 1$, one may expand exponentials to arrive at a simplified description.

6.2 Piecewise Constant Feedback Control

Even when a system is simply coupled to an equilibrium reservoir and rapidly equilibrates with that reservoir, an additional interaction with a detector may generally drive it out of equilibrium. In this section, we will discuss the propagator for the case when—immediately subsequent to the measurement—the performed control operation depends on the measurement result.

We have a setup in mind where a quantum system is monitored by a detector, which at regular time intervals Δt performs a measurement of some quantity. The measurement is assumed to be performed infinitely fast; i.e., there is no associated duration. If the evolution between the measurements does not depend on the measurement result, it can be described by a simple Kraus map, $\rho(t + \Delta t) = \sum_\alpha K_\alpha(\Delta t)\rho(t)K_\alpha^\dagger(\Delta t)$. Taking ρ as the density matrix right before the measurement, denoting m as the measurement outcome, and denoting ρ'_m as the density matrix right before the next measurement, these are then related by

$$\rho'_m = \sum_\alpha K_\alpha(\Delta t)\frac{M_m\rho M_m^\dagger}{\mathrm{Tr}\{M_m^\dagger M_m\rho\}}K_\alpha^\dagger(\Delta t). \qquad (6.3)$$

Thus, for a particular measurement outcome m, the density matrix at time $t + \Delta t$ depends non-linearly on the density matrix at time t. However, the probability of obtaining the measurement outcome m is given by $P_m = \mathrm{Tr}\{M_m^\dagger M_m\rho\}$, which means that by performing a weighted average over all outcomes we obtain

$$\rho' = \sum_m P_m\rho'_m = \sum_\alpha \sum_m K_\alpha(\Delta t)M_m\rho M_m^\dagger K_\alpha^\dagger(\Delta t)$$

$$\hat{=} \sum_\alpha \mathscr{K}_\alpha(\Delta t) \sum_m \mathscr{M}_m\rho \equiv \mathscr{K}(\Delta t)\mathscr{M}\rho. \qquad (6.4)$$

This equation includes only the evolution under measurement and the subsequent evolution and now constitutes a linear relation between ρ and ρ', which can also be written as a linear map or superoperator [10]. We note that $\mathscr{M} \neq \mathbf{1}$ expresses the fact that measurements may also modify the quantum system when the measurement result is discarded. Since the evolution after the measurement so far does not depend on the measurement result, the above equation would simply give rise to another open-loop control scheme, where measurement is used as a particular kind of control action. The averaged density matrix ρ' is the one which is relevant to calculate expectation values of observables,

$$\langle A \rangle = \sum_m P_m\, \mathrm{Tr}\{A\rho'_m\} = \mathrm{Tr}\{A\rho'\}, \qquad (6.5)$$

where we have just used linearity of the trace.

When now the evolution after the measurement depends on the measurement result, formally described by making the Kraus operators dependent on the result $K_\alpha(\Delta t) \to K_\alpha^{(m)}(\Delta t)$, we generate a feedback control loop, and the average density matrix after one cycle becomes

$$\rho' = \sum_\alpha \sum_m K_\alpha^{(m)}(\Delta t)M_m\rho M_m^\dagger K_\alpha^{(m)\dagger}(\Delta t)\hat{=} \sum_\alpha \sum_m \mathscr{K}_\alpha^{(m)}(\Delta t)\mathscr{M}_m\rho$$

$$= \mathscr{P}_{\mathrm{fb}}(\Delta t)\rho. \qquad (6.6)$$

This equation defines a propagator under the described feedback control scheme. When exactly the same sequence of measurement and conditioned control operation is applied over and over again, one arrives at the iterative scheme

$$\rho_{k+1} = \mathscr{P}_{\text{fb}}(\Delta t)\rho_k, \tag{6.7}$$

and it is now intriguing to investigate the properties—in a stroboscopic sense—of stationary states, which obey

$$\bar{\rho} = \mathscr{P}_{\text{fb}}(\Delta t)\bar{\rho}. \tag{6.8}$$

A particularly interesting special case arises when the conditioned evolution after the measurement can be expressed by a Lindblad generator,

$$\sum_{\alpha} \mathscr{K}_{\alpha}^{(m)} = \exp\{\mathscr{L}_m \Delta t\}, \tag{6.9}$$

which expresses the feedback propagator as

$$\mathscr{P}_{\text{fb}} = \sum_m e^{\mathscr{L}_m \Delta t} \mathscr{M}_m. \tag{6.10}$$

This immediately demonstrates that even with Lindblad master equations, additional measurement may have a significant impact on the evolution. When it is not compatible with the structure of the conditioned Liouvillian (e.g., when it does not obey the block structure of the conventional quantum optical master equation in Definition 2.3), the evolution under feedback control can be expected to differ strongly from the free evolution or also from that of open-loop control schemes. When furthermore the time interval between the measurements becomes very small, we may expand the exponential

$$\rho_{k+1} \approx \left[1 + \sum_m \mathscr{L}_m \mathscr{M}_m \Delta t + \mathscr{O}\{\Delta t\}^2\right]\rho_k, \tag{6.11}$$

which enables one to define an effective Liouvillian under continuous feedback control:

$$\mathscr{L}_{\text{fb}}\rho_k = \frac{\rho_{k+1} - \rho_k}{\Delta t} = \sum_m \mathscr{L}_m \mathscr{M}_m \rho_k \quad \Longrightarrow \quad \mathscr{L}_{\text{fb}} = \sum_m \mathscr{L}_m \mathscr{M}_m. \tag{6.12}$$

Even in the continuous control limit $\Delta t \to 0$ one thus concludes that if the conditioned Liouvillians $\mathscr{L}_m$ have certain properties such as a block structure in the system energy eigenbasis or simply entries in the population block obeying local detailed balance, this need no longer be true for the effective Liouvillian $\mathscr{L}_{\text{fb}}$. The measurement superoperators $\mathscr{M}_m$ may alter the structure of the resulting Liouvillian completely. This effect could for example be exploited to stabilize states that—without feedback control—would simply decay.

6.3 Wiseman–Milburn Quantum Feedback

By Wiseman–Milburn quantum feedback [3] we denote feedback control schemes in which only jumps in a quantum system are continuously monitored. In contrast to the previous section, the measurement is not performed directly on the system, but via the reservoir that triggers the quantum jumps. Since the detection of an emitted photon does not necessarily completely determine the state of the system that has emitted the photon, such measurements are also called weak. In a higher dimensional Hilbert space containing both the system and the detector, these measurements may however also be described as projective—a special manifestation of Neumark's theorem [11]. In Wiseman–Milburn feedback, upon detection of a quantum jump during the time interval Δt, one acts with a unitary control operation U_c that is performed infinitely fast and directly subsequent to the time interval Δt [12–14]. This type of feedback control scheme can be conveniently included in a master equation formalism: we consider a system with m states together with a detector monitoring transitions of the system (of which there are at most $D < m(m-1)$). The detector is not directly coupled to the system but only via a reservoir: it can, e.g., detect photons emitted from the system into the reservoir (vacuum). The system's evolution due to the reservoir is without explicit monitoring and feedback described by the Lindblad master equation

$$\dot{\rho} = -\mathrm{i}[H, \rho] + \sum_{\alpha=1}^{D} \gamma_\alpha \left[L_\alpha \rho L_\alpha^\dagger - \frac{1}{2}\{L_\alpha^\dagger L_\alpha, \rho\}\right], \tag{6.13}$$

where $\gamma_\alpha \geq 0$ represents the rate for the dimensionless Lindblad jump operator L_α. For demonstration, we assume that the detector in the reservoir detects only the jump corresponding to L_D. In quantum optics, the detection of a photon by a photomultiplier goes along with its destruction. Furthermore, photomultipliers cannot detect the inverse process, i.e., the absorption of a photon from the reservoir by the system. This may be different in an electronic context, where, e.g., the measurement of an electronic jump by a quantum point contact (QPC) does not destroy the electron, and also the detection of bidirectional electronic jumps is possible. To identify the measurement operators, we now discretize the master equation and map (similarly to the method in Sect. 1.5.2) to an iteration equation for the density matrix, which when $\Delta t \to 0$ exactly reproduces the Lindblad dynamics above:

$$\rho(t + \Delta t) = \sum_{\alpha\beta} w_{\alpha\beta}(\Delta t) K_\alpha \rho(t) K_\beta^\dagger \tag{6.14}$$

with the dimensionless operators

$$K_1 = \mathbf{1}, \qquad K_2 = \frac{-1}{2} \sum_\alpha \frac{\gamma_\alpha}{\gamma} L_\alpha^\dagger L_\alpha - \mathrm{i}H,$$

$$K_3 = L_1, \qquad \ldots, \qquad K_{D+2} = L_D, \tag{6.15}$$

where the overall decay rate $\gamma = \sum_{\alpha} \gamma_{\alpha}$ has been introduced to obtain dimensionless Kraus operators K_{α}. The dimensionless coefficient matrix is given by

$$
w = \begin{pmatrix}
1 & \gamma\,\Delta t & 0 & \cdots & 0 \\
\gamma\,\Delta t & 0 & 0 & \cdots & 0 \\
0 & 0 & \gamma_1\,\Delta t & & \\
\vdots & \vdots & & \ddots & \\
0 & 0 & & & \gamma_D\,\Delta t
\end{pmatrix}, \tag{6.16}
$$

which becomes a positivity-preserving map when $\Delta t \to 0$, i.e., in the master equation limit as shown in Sect. 1.5.2. Noting that contributions quadratic in Δt do not change the Lindblad master equation obtained when $\Delta t \to 0$, we may also modify $w_{22} \to (\gamma\,\Delta t)^2$, such that the map is also always positivity-preserving for finite Δt. This step would sacrifice trace preservation, which however may be restored by successive renormalization, yielding a nonlinear map-preserving hermiticity, trace, and positivity for finite $\Delta t > 0$.

Here, we will however be content with the continuous measurement limit. We choose to act with the unitary control operation U_c on the density matrix [3], when the jump L_D is detected,

$$
\rho \xrightarrow{D} U_c \rho U_c^{\dagger}, \tag{6.17}
$$

and to do nothing otherwise. To be consistent with the assumptions of Wiseman–Milburn feedback, we assume that performing the unitary operation is infinitely short. Realistically, the unitary control operation could be implemented by a δ-pulse-shaped Hamiltonian

$$
U_c = e^{-i \int_t^{t+\delta t} \alpha(t',\delta t)\,dt'\, H_c} = U_c^{\dagger}, \tag{6.18}
$$

where $\alpha(t,\delta t) = \alpha(t+\delta t,\delta t) = 0$ and $\int_t^{t+\delta t} \alpha(t',\delta t)\,dt' < \infty$ also when $\delta t \to 0$. The corresponding positivity-preserving map is obtained from Eq. (6.14) by modifying the evolution such that a control action is applied only when transition L_D is detected:

$$
\rho(t + \Delta t) = \sum_{\alpha\beta=1}^{D+1} w_{\alpha\beta}(\Delta t) K_{\alpha}\rho(t) K_{\beta}^{\dagger} + \gamma\,\Delta t\, U_c L_D \rho(t) L_D^{\dagger} U_c^{\dagger}. \tag{6.19}
$$

Due to the unitarity of the control operation $U_c^{\dagger} U_c = \mathbf{1}$, we can also obtain the above iteration map by simply replacing $L_D \to U_c L_D^{\dagger}$ throughout. Again using the limit $\Delta t \to 0$, we obtain the feedback-controlled master equation

$$
\dot{\rho} = -i[H, \rho] + \sum_{\alpha=1}^{D-1} \left[L_{\alpha}\rho L_{\alpha}^{\dagger} - \frac{1}{2}\{ L_{\alpha}^{\dagger} L_{\alpha}, \rho \} \right]
$$

$$
+ U_c L_D \rho L_D^{\dagger} U_c^{\dagger} - \frac{1}{2}\{ L_D^{\dagger} L_D, \rho \}. \tag{6.20}
$$

Comparing with Eq. (6.13), we see that we might have also obtained this equation by simply applying the unitary control operator right after the corresponding jump in the original master equation, which sometimes serves as a guiding principle of constructing feedback-controlled master equations. For multiple reservoirs, this can be directly extended to jump detection into a specific reservoir: then, only the jump corresponding to the particular reservoir has to be equipped with the unitary control operation. Furthermore, one may apply different controls for different jumps, such that Wiseman–Milburn feedback control also gives rise to a plethora of different evolution equations.

6.4   Further Roads to Feedback

The preceding examples of piecewise constant dissipative control or instantaneous unitary action are by no means exhaustive. One might also try to realize instantaneous dissipative control by modifying the tunneling rates of a single electron transistor in a pulse-like manner [15]. Alternatively, one could use measurements as an instantaneous nonunitary control action. Furthermore, the measurement basis does not have to coincide with the preferred basis induced by the reservoir, such that quite general iteration equations are conceivable.

To obtain a convenient representation of these approaches, we apply a superoperator notation. The evolution without measurement and feedback can be described by a positivity-preserving Kraus map:

$$\rho(t + \Delta t) = \sum_{\alpha} K_{\alpha}(\Delta t)\rho(t)K_{\alpha}^{\dagger}(\Delta t) \hat{=} \sum_{\alpha} \mathcal{K}_{\alpha}(\Delta t)\rho(t)$$

$$= \mathcal{K}(\Delta t)\rho(t). \tag{6.21}$$

The control operation can also be most generally described by a Kraus map $\mathcal{W}_c(\Delta\tau)$—whether it is unitary or not—where $\Delta\tau$ denotes the time necessary for performing the control. To keep the correspondence to Lindblad evolution, one could represent the more general control operations by an exponential

$$\mathcal{W}_c = e^{\kappa_c}, \tag{6.22}$$

where κ could be a time integral of a Liouvillian. However, to obtain a true feedback control scheme, the control action should be conditioned on the measurement result $\mathcal{W}_c \rightarrow \mathcal{W}_c^{\alpha}$. Even then, different ways of conditioning the control action are conceivable. For example, in the Wiseman–Milburn scheme, the measurement was performed during the time interval Δt, and if we conditioned the control action on the corresponding outcome, the evolution of the density matrix would be given by

$$\rho(t + \Delta t) = \sum_{\alpha} \mathcal{W}_c^{\alpha} \mathcal{K}_{\alpha}(\Delta t)\rho(t). \tag{6.23}$$

Alternatively, one might condition the control on the outcome of an instantaneous measurement—described by the measurement superoperators $\mathscr{M}_\alpha$ that occur right after the free evolution. This approach would yield the iteration equation

$$\rho(t + \Delta t) = \sum_\alpha \mathscr{W}_c^\alpha \mathscr{M}_\alpha \mathscr{K}(\Delta t)\rho(t). \tag{6.24}$$

More sophisticated schemes are conceivable.

References

1. L. Viola, E. Knill, S. Lloyd, Dynamical decoupling of open quantum systems. Phys. Rev. Lett. **82**, 2417 (1999)
2. S. Lloyd, L. Viola, Engineering quantum dynamics. Phys. Rev. A **65**, 010101 (2001)
3. H.M. Wiseman, G.J. Milburn, *Quantum Measurement and Control* (Cambridge University Press, Cambridge, 2010)
4. S. Juergens, F. Haupt, M. Moskalets, J. Splettstoesser, Thermoelectric performance of a driven double quantum dot. Phys. Rev. B **87**, 245423 (2013)
5. E. Barnes, S. Das Sarma, Analytically solvable driven time-dependent two-level quantum systems. Phys. Rev. Lett. **109**, 060401 (2012)
6. T. Brandes, F. Renzoni, R.H. Blick, Adiabatic steering and determination of dephasing rates in double-dot qubits. Phys. Rev. B **64**, 035319 (2001)
7. M.S. Sarandy, D.A. Lidar, Adiabatic approximation in open quantum systems. Phys. Rev. A **71**, 012331 (2005)
8. C.E. Creffield, Quantum control and entanglement using periodic driving fields. Phys. Rev. Lett. **99**, 110501 (2007)
9. I. Degani, A. Zanna, L. Saelen, R. Nepstad, Quantum control with piecewise constant control functions. SIAM J. Sci. Comput. **31**, 3566 (2009)
10. F. Ticozzi, L. Viola, Stabilizing entangled states with quasi-local quantum dynamical semigroups. Proc. R. Soc. Lond. Ser. A **370**, 5259 (2012)
11. A. Peres, Neumark's theorem and quantum inseparability. Found. Phys. **20**, 1441 (1990)
12. G. Kießlich, G. Schaller, C. Emary, T. Brandes, Charge qubit purification by an electronic feedback loop. Phys. Rev. Lett. **107**, 050501 (2011)
13. G. Kießlich, C. Emary, G. Schaller, T. Brandes, Reverse quantum state engineering using electronic feedback loops. New J. Phys. **14**, 123036 (2012)
14. C. Pöltl, C. Emary, T. Brandes, Feedback stabilization of pure states in quantum transport. Phys. Rev. B **84**, 085302 (2011)
15. G. Schaller, C. Emary, G. Kiesslich, T. Brandes, Probing the power of an electronic Maxwell's demon: single-electron transistor monitored by a quantum point contact. Phys. Rev. B **84**, 085418 (2011)

Chapter 7
Controlled Systems

Abstract This chapter discusses systems subjected to various forms of open-loop or closed-loop control. We begin with a single junction that is characterized only by two tunneling rates describing the probability for jumps in the respective direction. Whereas without feedback one simply obtains Poissonian statistics, modifying the tunneling rate either by open-loop protocols or closed-loop protocols may lead to a modified counting statistics of transferred particles. Here, only closed-loop protocols have the potential to induce significant changes. Second, we consider an electronic pump, where an open-loop protocol may transport electrons against a potential bias. It will be shown that this external control requires energy, without which the pump will not work. The situation is different in feedback control, where the information obtained from a measurement is used to choose the performed control action. The model thus implements an electronic version of Maxwell's demon: when the information current associated with the measurement is included in the entropy balance, the second law is respected. We also discuss an all-inclusive setup, where the demon is explicitly included in the dynamics. When treated as a whole, the device is nothing but a thermoelectric generator converting a temperature gradient into power. Finally, we outline the potential of feedback control to stabilize quantum coherence by discussing a single qubit that is periodically measured and connected to different reservoirs.

7.1 Single Junction

The simplest model for studying counting statistics is that of a single junction. Such a junction could be physically implemented by a quantum point contact (QPC),

$$H = \sum_k \varepsilon_{kL} c_{kL}^\dagger c_{kL} + \sum_k \varepsilon_{kR} c_{kR}^\dagger c_{kR} + \sum_{kk'} [t_{kk'} c_{kL} c_{k'R}^\dagger + t_{kk'}^* c_{k'R} c_{kL}^\dagger], \quad (7.1)$$

where $c_{k\alpha}$ are fermionic annihilation operators for electrons in mode k and lead α. The tunneling process from an electron of the left lead in mode k to the mode k' of the right lead is described by the term $t_{kk'} c_{kL} c_{k'R}^\dagger$, whereas the inverse process is described by the hermitian conjugate term. Altogether, the model is represented by a quadratic fermionic Hamiltonian and can therefore—with some effort—be solved

G. Schaller, *Open Quantum Systems Far from Equilibrium*, Lecture Notes in Physics 881,   159
DOI 10.1007/978-3-319-03877-3_7,
© Springer International Publishing Switzerland 2014

exactly (compare also Sect. 3.2). However, here we will rather be interested in the
effect of feedback, and will therefore be content with a perturbative treatment in the
tunneling amplitudes $t_{kk'}$. To obtain the statistics of charges traveling from left to
right, we append a virtual detector $B = \sum_n |n - 1\rangle\langle n|$ with infinitely many states
$|n\rangle$ to the system that increases its counter each time an electron is created on the
right lead. This modifies the interaction Hamiltonian to

$$H_I = B^\dagger \otimes \sum_{kk'} t_{kk'} c_{kL} c_{k'R}^\dagger + B \otimes \sum_{kk'} t_{kk'}^* c_{k'R} c_{kL}^\dagger, \tag{7.2}$$

which is thus represented in the conventional system-bath decomposition, where the
detector is now the system of interest. To obtain a rate equation as in Definition 2.4,
we evaluate the correlation function

$$C_{12}(\tau) = \frac{1}{2\pi} \int d\omega \, d\omega' \, T(\omega, \omega') f_R(\omega') [1 - f_L(\omega)] e^{-i(\omega-\omega')\tau},$$
$$C_{21}(\tau) = \frac{1}{2\pi} \int d\omega \, d\omega' \, T(\omega, \omega') f_L(\omega) [1 - f_R(\omega')] e^{+i(\omega-\omega')\tau}, \tag{7.3}$$

where the tunneling rate is defined as $T(\omega, \omega') = 2\pi \sum_{kk'} |t_{kk'}|^2 \delta(\omega - \varepsilon_{kL}) \delta(\omega - \varepsilon_{k'R})$, and their Fourier transforms therefore become

$$\gamma_{12}(\Omega) = \int d\omega \, T(\omega, \omega - \Omega) [1 - f_L(\omega)] f_R(\omega - \Omega),$$
$$\gamma_{21}(\Omega) = \int d\omega \, T(\omega, \omega + \Omega) f_L(\omega) [1 - f_R(\omega + \Omega)], \tag{7.4}$$

which can, e.g., for flat tunneling rates $T(\omega, \omega') \approx T_0$ be evaluated explicitly using
our results from Sect. 5.4. Assuming that the difference of the detector energies is
negligible, the transition rate from state n to $n + 1$ in the detector, i.e., the transition
rate of an electron transfer from left to right γ and the inverse transition rate $\bar{\gamma}$ are
then given by

$$\gamma = \gamma_{21}(0), \qquad \bar{\gamma} = \gamma_{12}(0). \tag{7.5}$$

Therefore, in this particular model—and many others—in the small tunneling limit
one will obtain the equations

$$\dot{P}_n = +\gamma P_{n-1}(t) + \bar{\gamma} P_{n+1} - [\gamma + \bar{\gamma}] P_n(t), \tag{7.6}$$

where $P_n(t)$ denotes the probability that n particles have passed the junction after
time t. That is, all the microscopic information contained in the tunneling amplitudes $t_{kk'}$ and the lead occupations $f_\alpha(\omega)$ is compressed only in the two tunneling
rates γ and $\bar{\gamma}$; see Fig. 7.1. Depending on the microscopic underlying model, these
parameters may depend on the particle concentrations on the left and right sides of
the tunneling barrier and on the height of the barrier, etc. Thus, they may be modified
in time by changing these microscopic parameters.

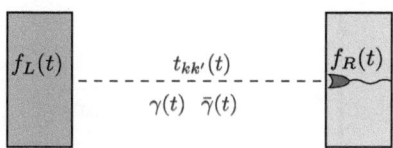

Fig. 7.1 Sketch of a single junction between two reservoirs, characterized by their Fermi functions f_α and tunneling amplitudes $t_{kk'}$. When the tunneling amplitudes are treated perturbatively, the time-dependent microscopic parameters just enter into the time-dependent left-to-right and right-to-left tunneling rates $\gamma(t)$ and $\bar{\gamma}(t)$, respectively. The piecewise constant time dependence may either follow a predefined protocol (open-loop control) or can be conditioned on a measurement result (feedback control). The system in this case is given by a virtual detector that counts the net number of particles transferred from left to right

First let us consider the time-independent case. After the Fourier transformation $P(\chi, t) = \sum_n P_n(t) e^{+in\chi}$, the n-resolved equation becomes

$$\dot{P}(\chi, t) = \left[\gamma\left(e^{+i\chi} - 1\right) + \bar{\gamma}\left(e^{-i\chi} - 1\right)\right] P(\chi, t). \tag{7.7}$$

This is thus in perfect agreement with what we had for the QPC statistics in Eq. (5.68). With the initial condition $P(\chi, 0) = 1$, it is solved by

$$P(\chi, t) = \exp\left\{\left[\gamma\left(e^{+i\chi} - 1\right) + \bar{\gamma}\left(e^{-i\chi} - 1\right)\right] t\right\}. \tag{7.8}$$

Exercise 7.1 (Cumulants) Show that the cumulants of the probability distribution $P_n(t)$ are given by

$$\langle\!\langle n^k \rangle\!\rangle = \left[\gamma + (-1)^k \bar{\gamma}\right] t,$$

and can thus be understood as two counter-propagating Poissonian distributions.

This initial condition is chosen because we assume that at time $t = 0$, no particle has crossed the junction $P_n(0) = \delta_{n,0}$. The probability of counting n particles after time t can be obtained from the inverse Fourier transform

$$P_n(t) = \frac{1}{2\pi} \int_{-\pi}^{+\pi} \exp\left\{\left[\gamma\left(e^{+i\chi} - 1\right) + \bar{\gamma}\left(e^{-i\chi} - 1\right)\right] t\right\} e^{-in\chi} \, d\chi. \tag{7.9}$$

This probability can be calculated analytically for this one-dimensional model even in the case of bidirectional transport:

$$P_n(t) = e^{-(\gamma + \bar{\gamma})t} \sum_{a,b=0}^{\infty} \frac{(\gamma t)^a}{a!} \frac{(\bar{\gamma} t)^b}{b!} \frac{1}{2\pi} \int_{-\pi}^{+\pi} e^{+i(a-b-n)\chi} d\chi$$

$$= e^{-(\gamma + \bar{\gamma})t} \sum_{a,b=0}^{\infty} \frac{(\gamma t)^a}{a!} \frac{(\bar{\gamma} t)^b}{b!} \delta_{a-b,n}$$

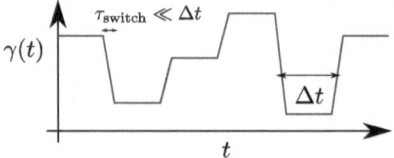

Fig. 7.2 Time-dependent tunneling rate which is (nearly) piecewise constant during the intervals Δt. In the model, we neglect the switching time τ_{switch} completely

$$= e^{-(\gamma+\bar{\gamma})t} \begin{cases} \sum_{a=n}^{\infty} \frac{(\gamma t)^a}{a!} \frac{(\bar{\gamma}t)^{a-n}}{(a-n)!}: & n \geq 0 \\ \sum_{a=0}^{\infty} \frac{(\gamma t)^a}{a!} \frac{(\bar{\gamma}t)^{a-n}}{(a-n)!}: & n < 0 \end{cases}$$

$$= e^{-(\gamma+\bar{\gamma})t} \left(\frac{\gamma}{\bar{\gamma}}\right)^{n/2} \mathscr{I}_n(2\sqrt{\gamma\bar{\gamma}}t), \tag{7.10}$$

where $\mathscr{I}_n(x)$ denotes a modified Bessel function of the first kind, defined as the solution of $z^2 \mathscr{I}_n''(z) + z \mathscr{I}_n'(z) - (z^2 + n^2) \mathscr{I}_n(z) = 0$. In the unidirectional transport limit, this reduces to a normal Poissonian distribution,

$$\lim_{\bar{\gamma}\to 0} P_n(t) = \begin{cases} e^{-\gamma t} \frac{(\gamma t)^n}{n!}: & n \geq 0, \\ 0: & n < 0. \end{cases} \tag{7.11}$$

Exercise 7.2 (Poissonian limit) Show that a Poissonian distribution arises in the unidirectional transport limit.

7.1.1 Open-Loop Control

Now we consider the case of a time-dependent rate $\gamma \to \gamma(t)$ with a piecewise constant time dependence. Just for simplicity, we will constrain ourselves to unidirectional transport $\bar{\gamma} = 0$ as shown in Fig. 7.2, where the time dependence of $\gamma(t)$ is well approximated by a piecewise constant protocol. We assume that the parameter γ is changed at regular time intervals Δt, such that the control protocol is fully characterized by the sequence $\{\gamma_1, \gamma_2, \ldots\}$. The fact that the model is scalar (has no internal structure) implies that the system has no internal memory, and the initial state for each interval is therefore simply that at which no particle has crossed the junction. Consequently, the probability distribution of measuring particles in the αth time interval is completely independent of the outcome of the interval $\alpha - 1$. If we denote the cumulant during the interval Δt in the αth interval by $\langle\!\langle n^k \rangle\!\rangle_\alpha$, we find for the average over all time intervals

$$\langle\!\langle \bar{n}^k \rangle\!\rangle = \frac{1}{N} \sum_{\alpha=1}^{N} \langle\!\langle n^k \rangle\!\rangle_\alpha = \frac{1}{N} \sum_{\alpha=1}^{N} \gamma_\alpha \Delta t = \langle\gamma\rangle \Delta t, \tag{7.12}$$

i.e., all average cumulants are simply described by the time-averaged tunneling rate. Regardless of the actual form of the protocol, one therefore always obtains a Poissonian distribution. In conclusion, piecewise constant open-loop control applied to a single junction will not substantially alter its dynamics.

Even if we had assumed bidirectional transport, this would not be different, as can be seen from Exercise 7.1.

7.1.2 Closed-Loop Control

For simplicity, we again consider the unidirectional transport limit, which is described by

$$\dot{\rho}_n = \gamma \rho_{n-1} - \gamma \rho_n. \tag{7.13}$$

The parameter γ describes the speed at which the resulting Poissonian distribution

$$\rho_n(\Delta t) = \begin{cases} e^{-\gamma \Delta t} \frac{(\gamma \Delta t)^n}{n!}: & n \geq 0, \\ 0: & n < 0 \end{cases} \tag{7.14}$$

moves towards larger n. This however also goes along with a spread of the distribution: its width $\sigma = \sqrt{\langle n^2 \rangle - \langle n \rangle^2}$ increases as $\sigma \propto t^{1/2}$. When we arrange the probabilities in an infinite-dimensional vector, the rate matrix appears in band-diagonal form:

$$\frac{d}{dt} \begin{pmatrix} \vdots \\ \rho_{n-1} \\ \rho_n \\ \vdots \end{pmatrix} = \begin{pmatrix} \ddots & & \\ \ddots & -\gamma & \\ & +\gamma & -\gamma \\ & & \ddots & \ddots \end{pmatrix} \begin{pmatrix} \vdots \\ \rho_{n-1} \\ \rho_n \\ \vdots \end{pmatrix}. \tag{7.15}$$

For the initial state $\rho_n(0) = \delta_{n,0}$ we have written the solution to the above equation explicitly in terms of a Poissonian distribution (7.14). Using the translational invariance in n and the linearity of the equations, we can therefore write the general solution explicitly as

$$\begin{pmatrix} \rho_0(t+\Delta t) \\ \rho_1(t+\Delta t) \\ \rho_2(t+\Delta t) \\ \vdots \\ \rho_n(t+\Delta t) \\ \vdots \end{pmatrix} = e^{-\gamma \Delta t} \begin{pmatrix} 1 & & & \\ \gamma \Delta t & 1 & & \\ \frac{(\gamma \Delta t)^2}{2} & \gamma \Delta t & 1 & \\ \vdots & \vdots & & \ddots & \ddots \\ \frac{(\gamma \Delta t)^n}{n!} & \frac{(\gamma \Delta t)^{n-1}}{(n-1)!} & \cdots & \cdots \\ \vdots & \vdots & & \end{pmatrix} \begin{pmatrix} \rho_0(t) \\ \rho_1(t) \\ \rho_2(t) \\ \vdots \\ \rho_n(t) \\ \vdots \end{pmatrix} \tag{7.16}$$

which takes the form $\rho(t + \Delta t) = \mathscr{W}(\Delta t)\rho(t)$ with the infinite-dimensional propagation matrix $\mathscr{W}(\Delta t)$.

Exercise 7.3 (Probability conservation) Show that the above introduced propagator $\mathscr{W}(\Delta t)$ preserves the sum of all probabilities, i.e., that $\sum_n \rho_n(t + \Delta t) = \sum_n \rho_n(t)$.

We have found previously that an open-loop control scheme does not drastically modify the probability distribution of tunneled particles. We now consider regular measurements of the number of tunneled particles being performed at time intervals Δt. The major difference from our previous considerations is that we now modify the tunneling rate γ dependent on the measured number of tunneled particles. The measurement of n tunneled particles can be described by a projective measurement of the density matrix. In superoperator notation, the matrix elements of the corresponding projector just read as

$$(\mathscr{P}_n)_{ij} = \delta_{i,n}\delta_{j,n}. \tag{7.17}$$

Conditioning the following propagator on the measurement result $\mathscr{W}(\Delta t) \to \mathscr{W}_n(\Delta t)$ via switching the tunneling rate depending on the measurement outcome, the effective propagator under feedback control (6.10) becomes

$$\mathscr{W}_{\text{fb}}(\Delta t) = \sum_n \mathscr{W}_n(\Delta t)\mathscr{P}_n. \tag{7.18}$$

Making everything explicit, the propagation matrix becomes

$$\mathscr{W}_{\text{fb}}(\Delta t) = \begin{pmatrix} e^{-\gamma_0 \Delta t} & & & \\ e^{-\gamma_0 \Delta t}(\gamma_0 \Delta t) & e^{-\gamma_1 \Delta t} & & \\ e^{-\gamma_0 \Delta t}\frac{(\gamma_0 \Delta t)^2}{2} & e^{-\gamma_1 \Delta t}(\gamma_1 \Delta t) & e^{-\gamma_2 \Delta t} & \\ \vdots & \vdots & \vdots & \ddots \\ e^{-\gamma_0 \Delta t}\frac{(\gamma_0 \Delta t)^n}{n!} & e^{-\gamma_1 \Delta t}\frac{(\gamma_1 \Delta t)^{n-1}}{(n-1)!} & e^{-\gamma_2 \Delta t}\frac{(\gamma_2 \Delta t)^{n-2}}{(n-2)!} & \cdots \\ \vdots & \vdots & \vdots & \ddots \end{pmatrix}. \tag{7.19}$$

The vector of probabilities under feedback evolves according to the iteration scheme $\rho(t + \Delta t) = \mathscr{W}_{\text{fb}}(\Delta t)\rho(t)$. Formally, every column thus corresponds to a different Poissonian process with tunneling rate γ_n.

Exercise 7.4 (Effective feedback propagator) Show the validity of Eq. (7.19).

The matrix elements of the effective feedback propagator thus read as

$$\mathscr{W}_{nm}(t, \Delta t) = \begin{cases} e^{-\gamma_m \Delta t}\frac{(\gamma_m \Delta t)^{(n-m)}}{(n-m)!} : & n \geq m, \\ 0 : & n < m. \end{cases} \tag{7.20}$$

The feedback protocol is now defined when one decides what action to perform in response to measuring a certain number of particles. For example, one may upon measurement of m total transferred particles at time t consider to change the tunneling rate for the next time interval to

$$\gamma_m(t) = \begin{cases} \gamma + \alpha(\gamma - \frac{m}{t}): & m \le \gamma t[1 + \frac{1}{\alpha}], \\ 0: & \text{else,} \end{cases} \tag{7.21}$$

where γ and α are feedback parameters. Stated in words, this choice would increase the tunneling rate during the next time interval when the measured particle number is smaller than γt and would decrease it if the measured particle number is larger than γt, keeping γ non-negative throughout. Such a feedback protocol would aim to stabilize a mean of $\langle m \rangle_t = \gamma t$. In addition however, one can also expect that the width of the resulting distribution will be modified, since trajectories that were "too slow" in one time interval will be accelerated in the next interval, whereas trajectories that are "too fast" will be slowed down.

In the following, we denote a general feedback protocol by the function $\gamma_n(t)$, which describes the tunneling rate in response to n total transferred particles measured at time t. Then, we obtain for the first moment at time $t + \Delta t$

$$\langle n \rangle_{t+\Delta t} = \sum_n n \rho_n(t + \Delta t) = \sum_n n \sum_m \mathcal{W}_{nm}(t, \Delta t) \rho_m(t)$$

$$= \sum_m \left[\sum_n n \mathcal{W}_{nm}(t, \Delta t) \right] \rho_m(t)$$

$$= \sum_m \left[\sum_{n=m}^{\infty} n e^{-\gamma_m \Delta t} \frac{(\gamma_m \Delta t)^{(n-m)}}{(n-m)!} \right] \rho_m(t)$$

$$= \sum_m \left[\sum_{n=0}^{\infty} (n+m) e^{-\gamma_m \Delta t} \frac{(\gamma_m \Delta t)^n}{n!} \right] \rho_m(t)$$

$$= \sum_m (\gamma_m \Delta t + m) \rho_m(t) = \langle \gamma_n \rangle_t \Delta t + \langle n \rangle_t, \tag{7.22}$$

such that the change of the first moment during time interval Δt is determined by the average tunneling rate. Similarly, we obtain for the second moment

$$\langle n^2 \rangle_{t+\Delta t} = \sum_n n^2 \rho_n(t + \Delta t) = \sum_m \left[\sum_{n=0}^{\infty} (n+m)^2 e^{-\gamma_m \Delta t} \frac{(\gamma_m \Delta t)^n}{n!} \right] \rho_m(t)$$

$$= \sum_m \left[\gamma_m \Delta t (1 + \gamma_m \Delta t) + 2\gamma_m \Delta t m + m^2 \right] \rho_m(t)$$

$$= \langle \gamma_n^2 \rangle_t \Delta t^2 + \langle \gamma_n(1 + 2n) \rangle_t \Delta t + \langle n^2 \rangle_t. \tag{7.23}$$

This implies for the variance under feedback

$$\langle n^2 \rangle_{t+\Delta t} - \langle n \rangle_{t+\Delta t}^2 = \Delta t^2 \big[\langle \gamma_n^2 \rangle_t - \langle \gamma_n \rangle_t^2 \big] + \Delta t \big[\langle \gamma_n \rangle_t + 2\langle \gamma_n n \rangle_t - 2\langle \gamma_n \rangle_t \langle n \rangle_t \big]$$
$$+ \langle n^2 \rangle_t - \langle n \rangle_t^2. \tag{7.24}$$

For large intervals Δt, the first term on the right-hand side will always dominate, such that due to its positivity, $\langle \gamma_n^2 \rangle_t - \langle \gamma_n \rangle_t^2 = \langle (\gamma_n - \langle \gamma_n \rangle)^2 \rangle \geq 0$, the variance will always increase during Δt. For small Δt however, the second term may dominate, and for adapted feedback protocols it may also become negative. Then, the variance of the distribution may decrease during Δt.

Exercise 7.5 (Variance evolution without feedback) Show that without feedback $\gamma_m(t) = \gamma$, the variance during the iteration will for arbitrary distributions always increase as $(\langle n^2 \rangle_{t+\Delta t} - \langle n \rangle_{t+\Delta t}^2) - (\langle n^2 \rangle_t - \langle n \rangle_t^2) = \gamma \Delta t$.

This feedback protocol would in some sense harness the stochasticity of the underlying process, as the resulting counting statistics would have a constant width but a mean increasing linearly in time. It is however clear that the resulting distribution cannot have a completely vanishing width, as for this case one would also obtain a growing variance in time.

Exercise 7.6 (Variance evolution of a localized distribution) Show that for arbitrary feedback, the variance of a localized distribution $\rho_m(t) = \delta_{m\bar{m}}$ will always increase unless $\gamma_{\bar{m}} = 0$.

Therefore, the resulting time-dependent distribution may be expected to have a finite width if the iteration scheme is performed in a range where the feedback is negative.

7.1.2.1 Exponential Feedback

Here, we would like to illustrate the effect of such a feedback protocol for the example of a continuous $(n \to x)$ Gaussian distribution

$$\rho_x(t) = \frac{1}{\sqrt{2\pi}\sigma} e^{-\frac{(x-\gamma t)^2}{2\sigma^2}} \tag{7.25}$$

at time t with mean $\mu = \gamma t = \langle x \rangle_t$ and width $\sigma = \sqrt{\langle x^2 \rangle_t - \langle x \rangle_t^2}$. Here, we consider the exponential feedback control scheme

$$\gamma_x(t) = \gamma \exp\left\{ \alpha(t) \left[1 - \frac{x}{\gamma t} \right] \right\} \tag{7.26}$$

with intended mean γ and $\alpha(t)$ representing a yet-to-be-determined control function. The expectation values (replacing sums by integrals and as an approximation extending the integration range over the complete real axis) can be readily computed:

$$\langle \gamma_x \rangle = \gamma e^{\frac{\alpha^2 \sigma^2}{2\gamma^2 t^2}}, \qquad \langle \gamma_x^2 \rangle = \gamma^2 e^{2\frac{\alpha^2 \sigma^2}{\gamma^2 t^2}}, \qquad \langle x \gamma_x \rangle = e^{\frac{\alpha^2 \sigma^2}{2\gamma^2 t^2}} \left[\gamma^2 t - \frac{\alpha \sigma^2}{t} \right]. \quad (7.27)$$

This implies a change of variance of

$$\Delta \sigma^2 = \Delta t^2 \big[\langle \gamma_x^2 \rangle_t - \langle \gamma_x \rangle_t^2 \big] + \Delta t \big[\langle \gamma_x \rangle_t + 2 \langle x \gamma_x \rangle_t - 2 \langle x \rangle_t \langle \gamma_x \rangle_t \big]$$

$$= \Delta t^2 \gamma^2 e^{\frac{\alpha^2 \sigma^2}{\gamma^2 t^2}} \big[e^{\frac{\alpha^2 \sigma^2}{\gamma^2 t^2}} - 1 \big] + \frac{\Delta t}{t} e^{\frac{\alpha^2 \sigma^2}{2\gamma^2 t^2}} \big[\gamma t - 2\alpha \sigma^2 \big]. \quad (7.28)$$

Neglecting the quadratic contribution (small Δt), the variance is stabilized when the feedback control function is adapted with the time and width of the distribution at time t

$$\alpha(t) = \frac{\gamma t}{2\sigma^2}. \quad (7.29)$$

Conversely, for a given feedback strength $\alpha = \alpha_0 t$, the above equation determines a stationary width of the distribution. This is also confirmed by the numerical propagation of an initially localized probability distribution with the iteration scheme (7.19) in Fig. 7.3. Without feedback (thin lines), the distribution propagates, but also increases its width in time. With a constant feedback control function α, the distribution propagates at the same speed and its width is reduced compared to the situation without feedback. Nevertheless, its width still increases with time at a small rate. Only when the feedback control function $\alpha(t) = \alpha_0 t$ is chosen to scale linearly in time, the shape of the distribution is stabilized. Given a feedback control function $\alpha(t) = \alpha_0 t$ with the protocol (7.26), the stationary width of the resulting distribution can be determined as follows: inserting this dependence in Eq. (7.28) leads to a condition under which the variance of a Gaussian distribution remains constant,

$$0 = (\gamma \Delta t) e^{y/2} \big[e^y - 1 \big] + 1 - 2 \frac{\gamma}{\alpha_0} y, \quad (7.30)$$

with the variable

$$y = \frac{\alpha_0^2 \sigma^2}{\gamma^2}. \quad (7.31)$$

This transcendental equation can for given Δt, γ, and α_0 be solved numerically for y, which in turn enables one to determine the stationary width σ. For example, for $\gamma \Delta t = 1$ and $\gamma/\alpha_0 = 100$ we find a numerical solution $y \approx 0.005025$, which corresponds to a width of $\sigma \approx 7.1$. This width is exactly found in the Gaussian fit of the numerical solution (compare the inset of Fig. 7.3). The validity of this effective description under feedback is further underlined by a direct sampling of the feedback protocol (also compare the inset): here, the feedback parameter is adjusted

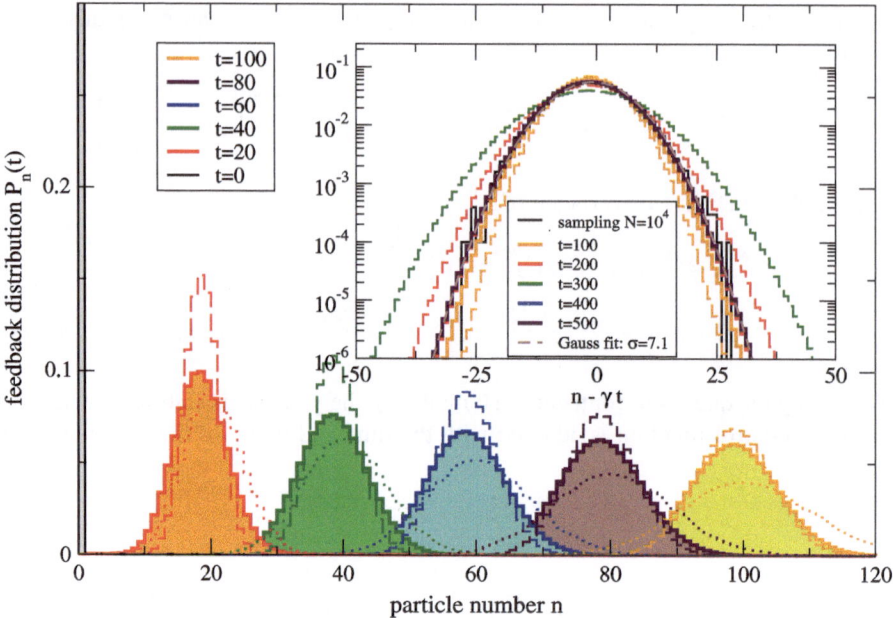

Fig. 7.3 Plot of probability distributions $P_n(t)$ without feedback (Poissonian evolution, *dotted*), with constant feedback strength (*dashed*), and with time-proportional feedback strength (*solid*). Without feedback, the width of the distribution will increase linearly in time (*dotted*). When the feedback parameter α is just constant, this can be reduced but not stopped (*dashed*). When α increases linearly in time, the width of the distribution is frozen while the mean increases (*solid*). The *inset* shows the long-time evolution for constant feedback strength (*dashed*, only up to $t = 300$) and time-proportional feedback strength (*solid*). The fit (*dashed curve*) demonstrates that the stationary distribution is Gaussian. In addition, a direct sampling of the feedback protocol yields the same distribution. Parameters: $\gamma = 1$, $\Delta t = 1$, $\alpha = 1$ for constant feedback and $\alpha = 0.01t$ for time-proportional feedback

at finite time intervals Δt, dependent on the previous measurement result. Therefore, under the chosen feedback protocol

$$\gamma_n(t) = \gamma \exp\left\{\alpha_0 t\left[1 - \frac{n}{\gamma t}\right]\right\}, \tag{7.32}$$

the long-term limit leads to a frozen distribution which propagates with a constant shape.

7.1.2.2 Linear Feedback

In contrast, linear feedback of the form

$$\gamma_n(t) = \gamma\left[1 - g(n - \gamma t)\right] \tag{7.33}$$

with the feedback parameters $g > 0$ and $\gamma > 0$ is much simpler to evaluate analytically as compared to the exponential protocol (7.26). It can be thought of as an approximation of scheme (7.32) for small $g = \alpha_0/\gamma \ll 1$. Of course, the above scheme formally allows for negative rates when $n \gg \gamma_0 t$. In reality however, the probability for such a process is exponentially suppressed for sufficiently large times, since for large times the width of a Poissonian process is sufficiently smaller than its mean value $\sigma/\mu = 1/\sqrt{\gamma t}$. The linear feedback scheme has the advantage that Eqs. (7.22) and (7.24) only couple to themselves and can thus be expressed only in terms of the first $C_1(t) = \langle n \rangle_t$ and second $C_2(t) = \langle n^2 \rangle_t - \langle n \rangle_t^2$ cumulants:

$$C_1(t + \Delta t) - C_1(t) = \gamma \Delta t \left[1 - g C_1(t) + g \gamma t \right],$$
$$C_2(t + \Delta t) - C_2(t) = \gamma^2 \Delta t^2 g^2 C_2(t) + \gamma \Delta t \left[1 + \gamma g t - g C_1(t) - 2g C_2(t) \right].$$
$$(7.34)$$

In the limit of small Δt, the differential version of these equations is

$$\dot{C}_1 = \gamma \left[1 - g C_1(t) + g \gamma t \right],$$
$$\dot{C}_2 = \gamma \left[1 + \gamma g t - g C_1(t) - 2g C_2(t) \right]$$
$$(7.35)$$

and admits for the initial conditions $C_1(0) = 0$ and $C_2(0) = 0$ the simple solution [1]

$$C_1(t) = \gamma t, \qquad C_2(t) = \frac{1 - e^{-2g\gamma t}}{2g},$$
$$(7.36)$$

which shows a continuous evolution towards a constant width of $\bar{\sigma} = \sqrt{\lim_{t \to \infty} C_2(t)} = \frac{1}{\sqrt{2g}}$.

Freezing the second cumulant of otherwise stochastic processes has many interesting applications. For example, many processes with a stochastic fluctuating work load might profit from a smoothed evolution if control can be applied. In an electronic context, a stabilized width of the electronic counting statistics could help to improve the standard of the electric current [2].

7.2 Electronic Pump

An electronic pump should perform work by transporting electrons against a potential gradient. As the simplest possible system we consider the single electron transistor (SET) from Sect. 5.1, which has just two internal states: empty and filled. The tunneling of electrons through the SET is stochastic and thereby in some sense uncontrolled. For example, the current fluctuations through such a device (noise) will increase linearly in time. A pump that regularly transports electrons would be expected to yield a large current with a small error rate, i.e., small noise. To remain in a picture where the concept of a pump makes sense, we allow only for time-dependent changes in the system and interaction Hamiltonians but keep the reservoirs invariant. See Fig. 7.4.

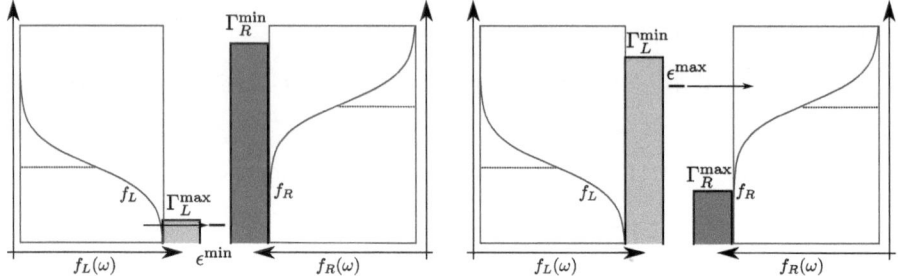

Fig. 7.4 Sketch of an SET subject to external open-loop control via the piecewise constant time-dependent dot level $\varepsilon(t)$ and the time-dependent tunneling rates $\Gamma_\alpha(t)$. The parameters of the reservoirs are assumed constant

Fig. 7.5 Sketch of two different configurations of dot level and tunneling rates with the same chemical potentials. When each setup is applied constantly, the average current will always flow from the lead with large chemical potential (*right*) to the lead with small chemical potential (*left*). In a transient regime however, an initially empty dot with a dot level shifted below both chemical potentials will—provided $\Gamma_L^{\max} \gg \Gamma_R^{\min}$—be dominantly filled from the left (*left panel*). If afterwards the dot level is lifted to a value larger than both chemical potentials and simultaneously the left barrier is raised and the right barrier is decreased, the dot will—provided $\Gamma_L^{\min} \ll \Gamma_R^{\max}$—dominantly unload towards the right lead (*right panel*). Effectively, this transfers an electron from left to right against the bias. Also when initially (*left panel*) the dot is filled, for the displayed protocol it is more likely that the electron is emitted to the right lead than to the left. Altogether, the scheme thus implements a stochastic electron pump

7.2.1 Power-Consuming Pump

Here, our aim is to keep the bias across the SET constant and to transport electrons against the bias by performing suitable open-loop control operations. Note that we use the term bias to describe the difference of lead occupations at the dot level; i.e., the bias can be induced by both a difference in chemical potentials or in the temperatures. We constrain ourselves to control operations that only alter the system or interaction Hamiltonian; i.e., the relative occupation of the reservoirs (which also defines the bias) should be left invariant. Therefore, we consider changing the dot level $\varepsilon(t)$ and the tunneling rates $\Gamma_\alpha(t)$—or the corresponding amplitudes $t_{k\alpha}(t)$ in the Hamiltonian—in a piecewise constant fashion. Intuitively, it is quite straightforward to arrive at a protocol that should lift electrons from left to right and thereby pump electrons from low to high chemical potentials; see Fig. 7.5. This just requires switching between the two possible configurations in the figure in constant time intervals. If the dot is initially empty, the most likely trajectory is as follows:

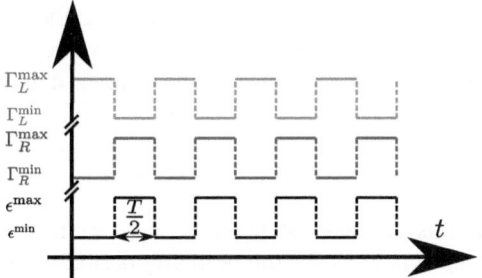

Fig. 7.6 Tunneling rates and the dot level are assumed to follow a periodic piecewise constant protocol, which admits only two possible values. The state of the contacts is assumed as stationary. For optimal cycle times T, the protocol is expected to pump electrons from the left towards the right contact (compare Fig. 7.5)

even though its occupation is lower than the right reservoir $f_L(\varepsilon_{\min}) < f_R(\varepsilon_{\min})$, the dot will be more likely loaded from the left reservoir, since the left bare tunneling rate is much larger than the right one, such that the overall rate of loading the pump from the left dominates, $\Gamma_L^{\max} f_L(\varepsilon_{\min}) \gg \Gamma_R^{\min} f_R(\varepsilon_{\min})$. After switching to the second configuration (right panel in Fig. 7.5), the dot will—although there is more space on the left than on the right $[1 - f_L(\varepsilon_{\max})] > [1 - f_R(\varepsilon_{\max})]$—dominantly unload towards the right reservoir, since the corresponding total rate is larger, $\Gamma_L^{\min}[1 - f_L(\varepsilon_{\max})] \ll \Gamma_R^{\max}[1 - f_R(\varepsilon_{\max})]$. In contrast, if the dot is initially filled, in the first step of the cycle it will most likely not unload, and in the second part of the cycle it will most likely unload to the right reservoir. Therefore, repeating these cycles periodically, one will expect the control scheme to act like a pump. Denoting the cycle time of the pump by T, we decide to apply each control configuration for time $T/2$, as displayed in Fig. 7.6. The rate matrices for the first and second half cycles therefore read

$$
\mathcal{L}_1(\chi) = \begin{pmatrix} -\Gamma_L^{\max} f_L(\varepsilon^{\min}) & +\Gamma_L^{\max}[1 - f_L(\varepsilon^{\min})] \\ -\Gamma_R^{\min} f_R(\varepsilon^{\min}) & +\Gamma_R^{\min}[1 - f_R(\varepsilon^{\min})]e^{+i\chi} \\ +\Gamma_L^{\max} f_L(\varepsilon^{\min}) & -\Gamma_L^{\max}[1 - f_L(\varepsilon^{\min})] \\ +\Gamma_R^{\min} f_R(\varepsilon^{\min})e^{-i\chi} & -\Gamma_R^{\min}[1 - f_R(\varepsilon^{\min})] \end{pmatrix},
$$

$$
\mathcal{L}_2(\chi) = \begin{pmatrix} -\Gamma_L^{\min} f_L(\varepsilon^{\max}) & +\Gamma_L^{\min}[1 - f_L(\varepsilon^{\max})] \\ -\Gamma_R^{\max} f_R(\varepsilon^{\max}) & +\Gamma_R^{\max}[1 - f_R(\varepsilon^{\max})]e^{+i\chi} \\ +\Gamma_L^{\min} f_L(\varepsilon^{\max}) & -\Gamma_L^{\min}[1 - f_L(\varepsilon^{\max})] \\ +\Gamma_R^{\max} f_R(\varepsilon^{\max})e^{-i\chi} & -\Gamma_R^{\max}[1 - f_R(\varepsilon^{\max})] \end{pmatrix},
$$

(7.37)

where we have already introduced counting fields to track the statistics of electrons transferred from left to right. The evolution of the density vector now follows the fixed-point iteration scheme

$$
\rho(t + T) = \mathcal{P}_2 \mathcal{P}_1 \rho(t) = e^{\mathcal{L}_2(0)T/2} e^{\mathcal{L}_1(0)T/2} \rho(t).
$$

(7.38)

After a period of transient evolution, the density vector at the end of the pump cycle will approach a value where $\rho(t+T) \approx \rho(t) = \bar{\rho}$. This value can—in a stroboscopic sense—be considered as a stationary density matrix and is defined by the eigenvalue equation

$$e^{\mathscr{L}_2(0)T/2} e^{\mathscr{L}_1(0)T/2} \bar{\rho} = \bar{\rho} \qquad (7.39)$$

with $\mathrm{Tr}\{\bar{\rho}\} = 1$. Once this stationary state has been determined, we can easily define the moment-generating functions for the pumping period in the (stroboscopically) stationary regime,

$$\mathscr{M}_1(\chi, T/2) = \mathrm{Tr}\{e^{\mathscr{L}_1(\chi)T/2} \bar{\rho}\},$$
$$\mathscr{M}_2(\chi, T/2) = \mathrm{Tr}\{e^{\mathscr{L}_2(\chi)T/2} e^{\mathscr{L}_1(0)T/2} \bar{\rho}\}, \qquad (7.40)$$

where $\mathscr{M}_1$ describes the statistics in the first half cycle and $\mathscr{M}_2$ in the second half cycle, and the moments can be extracted in the usual way. Since in the second half cycle, the initial state is not $\bar{\rho}$ but the propagator of the first half cycle applied to $\bar{\rho}$, we have modified the initial condition correspondingly. The distribution for the total number of particles tunneling during the complete pumping cycle $n = n_1 + n_2$ is given by

$$P_n(T) = \sum_{n_1,n_2:n_1+n_2=n} P_{n_1}(T/2) P_{n_2}(T/2)$$
$$= \sum_{n_1,n_2} \delta_{n,n_1+n_2} P_{n_1}(T/2) P_{n_2}(T/2). \qquad (7.41)$$

Then, the first moment may be calculated by simply adding the first moments of the particles tunneling through both half cycles:

$$\langle n \rangle = \sum_n n P_n = \sum_n n \sum_{n_1,n_2} \delta_{n,n_1+n_2} P_{n_1}(T/2) P_{n_2}(T/2)$$
$$= \sum_{n_1,n_2} (n_1 + n_2) P_{n_1}(T/2) P_{n_2}(T/2)$$
$$= \langle n_1 \rangle + \langle n_2 \rangle. \qquad (7.42)$$

The second cumulant can be similarly calculated:

$$\langle n^2 \rangle - \langle n \rangle^2 = \langle n_1^2 \rangle - \langle n_1 \rangle^2 + \langle n_2^2 \rangle - \langle n_2 \rangle^2. \qquad (7.43)$$

Exercise 7.7 (Second cumulant for joint distributions) Show the validity of Eq. (7.43).

This is just a manifestation of the fact that cumulants of independent stochastic processes are additive. In the following, we will constrain ourselves for simplicity to symmetric tunneling rates left and right,

$$\Gamma_L^{min} = \Gamma_R^{min} = \Gamma_{min}, \qquad \Gamma_L^{max} = \Gamma_R^{max} = \Gamma_{max}. \qquad (7.44)$$

Then, the long-term probability of finding the dot filled at the end of the pumping cycle becomes

$$\bar{P}_F = \frac{1}{\Gamma(1 + e^{-\Gamma T/2})} \left[\Gamma_{min} f_L(\varepsilon_{max}) + \Gamma_{max} f_R(\varepsilon_{max}) \right]$$

$$+ \frac{1}{\Gamma(1 + e^{+\Gamma T/2})} \left[\Gamma_{max} f_L(\varepsilon_{min}) + \Gamma_{min} f_R(\varepsilon_{min}) \right], \qquad (7.45)$$

where $\Gamma = \Gamma_{min} + \Gamma_{max}$ and the other probability is just $\bar{P}_E = 1 - \bar{P}_F$. With this stroboscopic long-term occupation, we obtain for the average number of electrons transferred to the right junction during one pumping cycle the expression

$$\langle n \rangle = \frac{\Gamma_{max}\Gamma_{min}}{\Gamma} \frac{T}{2} \left[f_L(\varepsilon^{max}) - f_R(\varepsilon^{max}) + f_L(\varepsilon^{min}) - f_R(\varepsilon^{min}) \right]$$

$$+ \frac{\Gamma_{max} - \Gamma_{min}}{\Gamma^2} \left[\Gamma_{max}(f_L(\varepsilon^{min}) - f_R(\varepsilon^{max})) \right.$$

$$\left. + \Gamma_{min}(f_R(\varepsilon^{min}) - f_L(\varepsilon^{max})) \right] \tanh[\Gamma T/4]. \qquad (7.46)$$

The first term (which dominates for slow pumping, i.e., large T) is negative when $f_L(\omega) < f_R(\omega)$, which at equal temperatures is the case when $\mu_L < \mu_R$. It is simply given by the average of the two SET currents one would obtain for the two half cycles. The second term however is present when $\Gamma_{max} > \Gamma_{min}$ and may be positive when the dot level is changed strongly enough such that $f_L(\varepsilon^{min}) > f_R(\varepsilon^{max})$ and $f_R(\varepsilon^{min}) > f_L(\varepsilon^{max})$. For large differences in the tunneling rates and small pumping times T, it may dominate the first term, such that the net particle number may be positive, even though the bias would normally favor the opposite current direction. When the tunneling rates are extremely different, $\Gamma_{min}T \to 0$, $\Gamma_{max}T \to \infty$, and also the dot level is strongly modified such that the Fermi functions obey $f_\alpha(\varepsilon^{min}) \approx 1$ and $f_\alpha(\varepsilon^{max}) \approx 0$, the average number per cycle approaches one. The noise can be calculated similarly (not shown for brevity).

It should be kept in mind that the realized electronic pump requires energy to work: by shifting the dot level from ε_{min} to ε_{max} we insert the energy $\Delta E = \varepsilon_{max} - \varepsilon_{min}$ in every pump cycle with a loaded dot after the first half into the system. When power consumption of the pump is not a problem (e.g., when the energy supply is unlimited), we may shift the dot level to very low and very large values, respectively. Then, the bias becomes negligible, and we can approximate $f_L(\varepsilon^{min}) \approx f_R(\varepsilon^{min}) \approx 1$ and $f_L(\varepsilon^{max}) \approx f_R(\varepsilon^{max}) \approx 0$. The particle number per cycle simplifies to

$$\langle n \rangle = \frac{\Gamma_{max} - \Gamma_{min}}{\Gamma_{max} + \Gamma_{min}} \tanh\left[\frac{T}{4}(\Gamma_{min} + \Gamma_{max})\right], \qquad (7.47)$$

and the noise $\langle n^2 \rangle - \langle n \rangle^2$ becomes ($\Gamma = \Gamma_{min} + \Gamma_{max}$)

$$\langle n^2 \rangle - \langle n \rangle^2 = 2 \frac{e^{\Gamma T/2} - 1}{(e^{\Gamma T/2} + 1)^2} \frac{\Gamma_{min}^2 + \Gamma_{max}^2 + (1 + e^{\Gamma T/2})\Gamma_{min}\Gamma_{max}}{\Gamma^2}. \quad (7.48)$$

For slow pumping, we therefore obtain for the Fano factor in this limit

$$F = \frac{\langle n^2 \rangle - \langle n \rangle^2}{|\langle n \rangle|} \xrightarrow{T \to \infty} \frac{2\Gamma_{max}\Gamma_{min}}{\Gamma_{max}^2 - \Gamma_{min}^2}, \quad (7.49)$$

which demonstrates that the pump works efficiently and noiselessly when, in addition to the extremely changed dot level, $\Gamma_{max} \gg \Gamma_{min}$. In this extreme limit the pump will act as a reliable single-electron source. Unfortunately, the pump has to run comparably slow to remain in this noiseless limit.

Furthermore, we note that at infinite bias, $f_L(\varepsilon_{min}) = f_L(\varepsilon_{max}) = 1$ and $f_R(\varepsilon_{min}) = f_R(\varepsilon_{max}) = 0$, the pump does not transport electrons against any potential or thermal gradient but can still be used to control the statistics of the tunneled electrons. For example, it is possible to also reduce the noise to zero in this limit when $\Gamma_{min} \to 0$.

To calculate the efficiency of the pump at finite bias and finite cycle time, we have to relate the work performed per pump cycle, $W_{out} = \langle n \rangle (\mu_R - \mu_L)$, to the energy consumed by the pump during switching from the first half cycle to the second, $W_{in} = \bar{n}_{1/2}(\varepsilon_{max} - \varepsilon_{min})$, where

$$\bar{n}_{1/2} = \text{Tr}\{\mathcal{N}_d e^{\mathcal{L}_1(0)T/2} \bar{\rho}\}$$

$$= \frac{1}{\Gamma(1 + e^{\Gamma T/2})}[\Gamma_{min} f_L(\varepsilon_{max}) + \Gamma_{max} f_R(\varepsilon_{max})]$$

$$+ \frac{1}{\Gamma(1 + e^{-\Gamma T/2})}[\Gamma_{max} f_L(\varepsilon_{min}) + \Gamma_{min} f_R(\varepsilon_{min})] \quad (7.50)$$

denotes the average occupation after the first half cycle. Figure 7.7 displays the average charge transport from left to right per cycle $\langle n \rangle$ (which by itself can be seen as a pump efficiency), the performed dimensionless work $\beta(\mu_R - \mu_L)\langle n \rangle$, and the energetic efficiency

$$\eta = \frac{\langle n \rangle \beta(\mu_R - \mu_L)}{\bar{n}_{1/2}\beta(\varepsilon_{max} - \varepsilon_{min})} \quad (7.51)$$

versus the pump cycle time T. The figure demonstrates that the efficiency—as the current output—is maximal at intermediate cycle times T.

7.2.2 Open-Loop Control at Zero Power Consumption

It would be intriguing to have a protocol that performs the same task but does not consume power, i.e., one that only changes the tunneling rates $\Gamma_\alpha(t)$. Unfortunately,

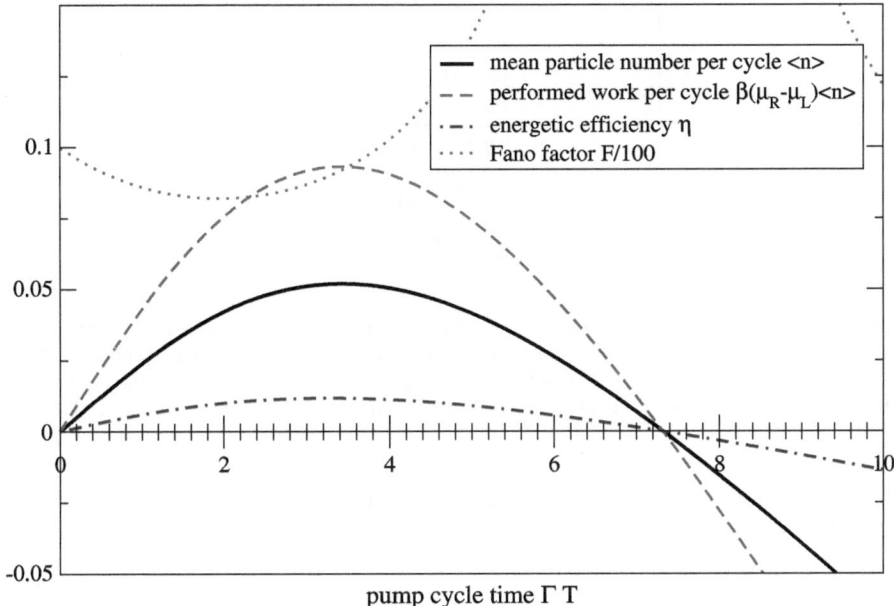

Fig. 7.7 Average number of tunneled particles (*solid black*), average work performed (*dashed red*), and energetic efficiency (*dash–dotted blue*) during one pump cycle versus cycle time ΓT. The rescaled Fano factor $F = (\langle n^2 \rangle - \langle n \rangle^2)/|\langle n \rangle|$ is also shown (*dotted green*). Since a constant bias is applied, the performed work per cycle is just proportional to the current, and for the chosen parameters the energetic efficiency of the pump is fairly small. For pumping that is too fast (small ΓT), the current vanishes since the pump cannot load and unload, whereas for pumping that is too slow (large ΓT), the current becomes negative, since the pump is too slow to overcome the applied bias. Parameters have been chosen as $\Gamma_{\max} = 9\Gamma_{\min}, f_L(\varepsilon^{\min}) = 0.6$, $f_R(\varepsilon^{\min}) = 0.9$, $f_L(\varepsilon^{\max}) = 1 - f_R(\varepsilon_{\min})$, $f_R(\varepsilon^{\max}) = 1 - f_L(\varepsilon_{\min})$

this cannot be expected to work since this would generate a perpetuum mobile. For consistency, we show here that by just assuming piecewise constant time dependencies of the tunneling rates $\Gamma_L(t)$ and $\Gamma_R(t)$ turned on in an arbitrary protocol while the dot level is kept constant, it is not possible to obtain an average current against the bias. The intended situation is also depicted in Fig. 7.8. Let the tunneling rates during the ith time interval be denoted by Γ_L^i and Γ_R^i and the corresponding constant Liouvillian during this time interval by $\mathscr{L}_i(\chi)$:

$$\mathscr{L}_i(\chi) = \begin{pmatrix} -\Gamma_L^i f_L - \Gamma_R^i f_R & +\Gamma_L^i(1 - f_L) + \Gamma_R^i(1 - f_R)e^{+i\chi} \\ +\Gamma_L^i f_L + \Gamma_R^i f_R e^{-i\chi} & -\Gamma_L^i(1 - f_L) - \Gamma_R^i(1 - f_R) \end{pmatrix}. \quad (7.52)$$

The density matrix after the time interval Δt is now given by

$$\rho_{i+1} = e^{\mathscr{L}_i(0)\Delta t} \rho_i, \quad (7.53)$$

such that it no longer has the form of a master equation but becomes a fixed-point iteration. In what follows, we will assume $f_L \leq f_R$ without loss of generality, such

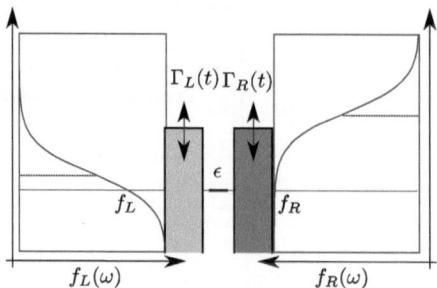

Fig. 7.8 Time-dependent tunneling rates which follow a piecewise constant protocol. An open-loop protocol is assumed; i.e., the control action does not depend on the occupation of the dot. Both the dot level and left and right Fermi functions are assumed constant in order not to inject energy into the system and to maintain the bias, respectively

that the source lead is right and the drain lead is left. The stationary state of the Liouvillian $\mathscr{L}_i$ is given by $(1 - \bar{f}_i, \bar{f}_i)$ with the stationary occupation

$$\bar{f}_i = \frac{\Gamma_L^i f_L + \Gamma_R^i f_R}{\Gamma_L^i + \Gamma_R^i}. \tag{7.54}$$

Since we have $\Gamma_{L/R}^i \geq 0$, this is just a convex combination, such that one has $f_L \leq \bar{f}_i \leq f_R$. This implies that after a few transient iterations and independent of the control protocol chosen, the density vector $\rho_i = (1 - f_i, f_i)$ will hover around, but will always remain inside the transport window constrained by f_L and f_R:

$$f_L \leq f_i = \langle d^\dagger d \rangle_i \leq f_R \quad \forall i \geq i^*. \tag{7.55}$$

The occupations will follow the iteration equation

$$\begin{pmatrix} 1 - f_{i+1} \\ f_{i+1} \end{pmatrix} = e^{\mathscr{L}_i(0)\Delta t} \begin{pmatrix} 1 - f_i \\ f_i \end{pmatrix}. \tag{7.56}$$

To calculate the number of particles tunneling into the right reservoir during the ith time interval, we consider the moment-generating function

$$\mathscr{M}_i(\chi, \Delta t) = \mathrm{Tr}\left\{ e^{\mathscr{L}_i(\chi)\Delta t} \begin{pmatrix} 1 - f_i \\ f_i \end{pmatrix} \right\} \tag{7.57}$$

and calculate the first moment

$$\langle n \rangle_i = (-\mathrm{i})\partial_\chi \mathscr{M}_i(\chi, \Delta t)|_{\chi=0}$$

$$= \frac{1}{(\Gamma_L^i + \Gamma_R^i)^2} \left[-\Gamma_R^2 (f_R - f_i)\left(1 - e^{-(\Gamma_L^i + \Gamma_R^i)\Delta t}\right) \right.$$

$$+ \Gamma_L^i \Gamma_R^i \big((f_i - f_L)\big(1 - e^{-(\Gamma_L^i + \Gamma_R^i)\Delta t}\big)$$

$$- (f_R - f_L)\big(\Gamma_L^i + \Gamma_R^i\big)\Delta t\big)\big] \le 0. \tag{7.58}$$

The first term is evidently negative when f_i is within the transport window $f_L \le f_i \le f_R$. The second term is also negative, which follows from its upper bound $f_i \to f_R$: here, we can use that $1 - e^{-x} \le x$ for all $x \ge 0$ with $x = (\Gamma_L^i + \Gamma_R^i)\Delta t$. Since the first moment is negative, $\langle n \rangle_i \le 0$ for all sufficiently large i, we conclude that the average number of particles after N iterations is also negative, $\langle n \rangle = \frac{1}{N}\sum_{i=1}^{N}\langle n \rangle_i \le 0$, provided N is sufficiently large. Actually, by choosing infinitesimally small time intervals $\Delta t \to 0$, arbitrary time dependencies $\Gamma_\alpha(t)$ can be approximated, such that the conclusion holds for any open-loop control protocol that modifies only the tunneling rates.

It follows that the average current always points from right (source) to left (drain) regardless of the actual protocol $\Gamma_{L/R}^\alpha(t)$ chosen. In other words, by simply modifying the tunneling rates (not taking into account whether the dot is occupied or not) it is not possible to revert the direction of the current. This result is not unexpected and is consistent with thermodynamics. However, it will be shown in the following sections that feedback control can be used to drive a current against the bias, when only the tunneling rates are adapted in a conditioned fashion.

7.3 Encoding Maxwell's Demon as Feedback Control

Originally, Maxwell's demon was introduced by Maxwell to underline the macroscopic character of thermodynamics [3]. Maxwell's demon is an intelligent being closely monitoring molecules moving inside a box. A wall partitions the box into two components, and by using a shutter, the demon can open a hole in the wall and thereby allow molecules to pass through the wall. Initially, the gas in both parts of the box is in thermal equilibrium, which may be achieved by opening the shutter for a sufficiently long time. Now, the demon performs the following feedback control protocol: when a fast molecule approaches the wall from the left, the shutter will open and the molecule will be transferred to the right compartment. When a slow molecule approaches the wall from the left, the shutter will close, and the molecule will remain in the left compartment. Molecules approaching the wall from the right are treated in the opposite fashion; only the slow ones are allowed to pass. After some time, the fast molecules will be on the right side and the slow ones on the left, such that the entropy of the box has decreased. The resulting temperature gradient can be used to generate useful work, which is surprising, because ideally opening or closing the shutter does not require energy. The common resolution to the paradox, known as Landauer's principle [4], is that the missing entropy is generated in the demon's brain during deletion. This also goes along with the dissipation of heat by the demon.

With our increasing abilities to control and manipulate the smallest systems, it has now become possible to implement such systems in the lab [5–7]. With these

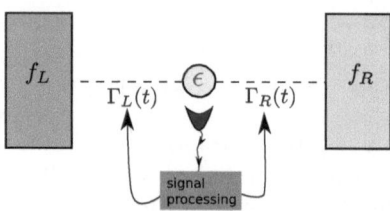

Fig. 7.9 Sketch of an SET subject to feedback control. A suitable (backaction-free as in Sect. 5.4) detector such as a quantum point contact monitors the time-dependent occupation of the central quantum dot. The detector signal is classically processed by an external device, which acts like Maxwell's demon. As control actions, the bare tunneling rates $\Gamma_\alpha(t)$ are changed in a piecewise constant fashion and conditioned on the detector signal. To inject no energy directly into the system, the dot level ε must remain constant

explicit models, it is possible to improve our understanding of how information can be converted into useful power.

7.3.1 Feedback Control Loop

Here, consider the simple example of an SET additionally coupled to a suitable charge detector such as, e.g., a QPC. The current through the QPC will as a macroscopic variable be read out, which leads to a projection of the SET dot onto empty or filled, respectively. Thus, in an idealized limit where the QPC tunneling rates are much larger than the dot tunneling rates, the measurement superoperators simply read as (we order the density matrix as $\rho_{EE}, \rho_{FF}, \rho_{EF}, \rho_{FE}$ in a vector)

$$
\mathcal{M}_E = \begin{pmatrix} 1 & 0 & 0 & 0 \\ 0 & 0 & 0 & 0 \\ 0 & 0 & 0 & 0 \\ 0 & 0 & 0 & 0 \end{pmatrix}, \qquad \mathcal{M}_F = \begin{pmatrix} 0 & 0 & 0 & 0 \\ 0 & 1 & 0 & 0 \\ 0 & 0 & 0 & 0 \\ 0 & 0 & 0 & 0 \end{pmatrix}, \qquad (7.59)
$$

which respects the block structure separating populations (the first two entries) and coherences (other entries). Here, we have neglected the physical backaction of the detector on the probed system. That is, just as in Sect. 5.4, the system is unaffected by the detector as long as we do not use the detector signal to change the system parameters.

Now in response to the measurement result (large current means an empty SET and low QPC current a filled SET, respectively), one changes the tunneling rates of the SET by rapidly changing the voltage of nearby gates. Normally, this will also lead to a shift of the electronic dot level, but a clever experimentalist may be able to adjust the potential landscape such that the dot level ε remains constant under such transformations, as displayed in Fig. 7.9. If this is done sufficiently fast, all tunneling rates become conditioned on the measurement outcome, formally described by replacing $\Gamma_\alpha(\varepsilon) \rightarrow \Gamma_\alpha^{E/F}$. For generality, we here also replace $f_\alpha(\varepsilon) \rightarrow f_\alpha^{E/F}$

(which holds when it is not possible to keep the dot level constant). The population parts of the corresponding conditioned Liouvillians thus read as

$$
\mathscr{L}_E = \begin{pmatrix} -\Gamma_L^E f_L^E - \Gamma_R^E f_R^E & +\Gamma_L^E(1-f_L^E) + \Gamma_R^E(1-f_R^E) \\ -\Gamma_L^E f_L^E - \Gamma_R^E f_R^E & +\Gamma_L^E(1-f_L^E) + \Gamma_R^E(1-f_R^E) \end{pmatrix},
$$
$$
\mathscr{L}_F = \begin{pmatrix} -\Gamma_L^F f_L^F - \Gamma_R^F f_R^F & +\Gamma_L^F(1-f_L^F) + \Gamma_R^F(1-f_R^F) \\ -\Gamma_L^F f_L^F - \Gamma_R^F f_R^F & +\Gamma_L^F(1-f_L^F) + \Gamma_R^F(1-f_R^F) \end{pmatrix},
$$
(7.60)

and it is obvious that each rate matrix separately obeys local detailed balance. For equal temperatures $\beta_L = \beta_R$, this implies that if one constantly uses either the tunneling rates Γ_α^E or Γ_α^F, the current will on average always flow from the junction with high chemical potential towards the junction with low chemical potential, regardless of the value of the tunneling rates.

Since the measurement superoperators obey the block structure separating coherences and populations (the underlying measurement operators commute with the system Hamiltonian), the effective Liouvillian under feedback will also decouple coherences and populations. In particular, its population part becomes

$$
\mathscr{L}_{\mathrm{fb}} \begin{pmatrix} -\Gamma_L^E f_L^E - \Gamma_R^E f_L^E & +\Gamma_L^F(1-f_L^F) + \Gamma_R^F(1-f_R^F) \\ -\Gamma_L^E f_L^E - \Gamma_R^E f_R^E & +\Gamma_L^F(1-f_L^F) + \Gamma_R^F(1-f_R^F) \end{pmatrix},
$$
(7.61)

and now no longer respects local detailed balance. In particular, even if the control loop is engineered in such a way that the dot level is not modified as the tunneling amplitudes in the Hamiltonian are changed ($f_\alpha^{(E/F)} \to f_\alpha$), we would obtain a modification of local detailed balance,

$$
\ln \frac{\mathscr{L}_{EF}^{(\alpha)}}{\mathscr{L}_{FE}^{(\alpha)}} = \ln \frac{\Gamma_\alpha^F}{\Gamma_\alpha^E} + \beta_\alpha(\varepsilon - \mu_\alpha) \equiv \delta_{FE}^{(\alpha)} + \beta_\alpha(\varepsilon - \mu_\alpha).
$$
(7.62)

Below, it will be demonstrated that this may have strong physical consequences.

7.3.2 Current

For example, with sufficiently modified tunneling rates it is now possible to obtain on average a current at equilibrium or even against an applied potential bias. Qualitatively, this behavior can be expected by intuitively considering the extremal scenario: when the dot is occupied, we might close the left barrier $\Gamma_L^E \to 0$, and when it is empty, we might close the right barrier $\Gamma_R^F \to 0$, whereas both Γ_L^F and Γ_R^E remain finite. This will clearly rectify the current from left to right. A more quantitative analysis can be extracted from the counting-field-dependent Liouvillian:

$$
\mathscr{L}_{\mathrm{fb}}(\chi) = \begin{pmatrix} -\Gamma_L^E f_L - \Gamma_R^E f_L & +\Gamma_L^F (1 - f_L) e^{-i\chi} + \Gamma_R^F (1 - f_R) \\ -\Gamma_L^E f_L e^{+i\chi} - \Gamma_R^E f_R & +\Gamma_L^F (1 - f_L) + \Gamma_R^F (1 - f_R) \end{pmatrix}. \quad (7.63)
$$

The associated matter current from left to right becomes

$$
I_M^L = \frac{\Gamma_L^E \Gamma_R^F f_L (1 - f_R) - \Gamma_L^F \Gamma_R^E (1 - f_L) f_R}{\Gamma_L^E f_L + \Gamma_L^F (1 - f_L) + \Gamma_R^E f_R + \Gamma_R^F (1 - f_R)}, \quad (7.64)
$$

which does not vanish at equilibrium $f_L = f_R$. For consistency, we note that the normal SET current is reproduced when feedback is turned off, $\Gamma_\alpha^{E/F} \to \Gamma_\alpha$.

At finite bias, the current may be directed against the bias, and thus the device may generate useful power just by using information. For simplicity, we parametrize the tunneling rates by a single parameter:

$$
\Gamma_L^E = \Gamma \frac{e^{+\delta}}{\cosh(\delta)}, \qquad \Gamma_L^F = \Gamma \frac{e^{-\delta}}{\cosh(\delta)},
$$

$$
\Gamma_R^E = \Gamma \frac{e^{-\delta}}{\cosh(\delta)}, \qquad \Gamma_R^F = \Gamma \frac{e^{+\delta}}{\cosh(\delta)}, \quad (7.65)
$$

where $\delta = 0$ reproduces the case of symmetric tunneling rates $\Gamma_L = \Gamma_R$ without feedback. In contrast, positive feedback strength $\delta > 0$ favors transport from left to right, which becomes unidirectional when $\delta \to \infty$; see also Fig. 7.10. The figure shows that the device may transport electrons against the bias, thereby generating power $P = -I_M^L (\mu_L - \mu_R)$. To obtain the point at which the power output is maximal, we maximize at infinite feedback strength $\delta \to \infty$ and under the assumption $f = f_L = 1 - f_R$ the expression

$$
\frac{\beta}{\Gamma} P = -\frac{I_M^L}{\Gamma} \beta (\mu_L - \mu_R) = -\left(\frac{2 f_L (1 - f_R)}{f_L + 1 - f_R} \right) \ln \left(\frac{f_L (1 - f_R)}{(1 - f_L) f_R} \right)
$$

$$
= -f \ln \left(\frac{f^2}{(1 - f)^2} \right) \quad (7.66)
$$

numerically with respect to f. Eventually, one obtains for $f = f_L = 1 - f_R \approx 0.2178$ at maximum feedback strength $\delta \to \infty$ a maximum power output of

$$
P_{\max} \approx \frac{\Gamma}{\beta} 0.5569, \quad (7.67)
$$

which for $\varepsilon = 0$ would be achieved at bias voltage $\beta V_{\max} \approx -2.5571$.

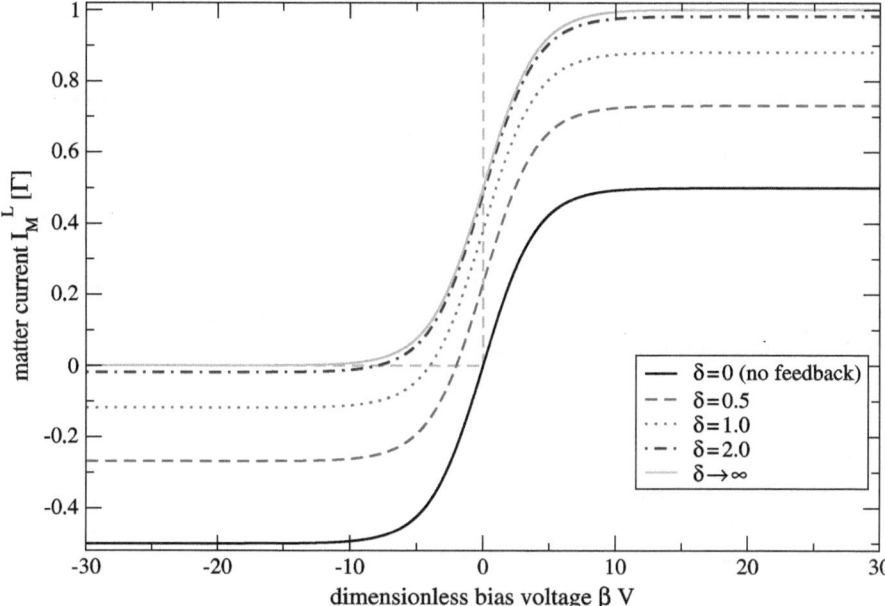

Fig. 7.10 Plot of the matter current versus dimensionless bias voltage for vanishing dot level $\varepsilon = 0$. Without feedback, the current is always directed with the bias (*solid black*). With increasing feedback strength, the transport from left to right is favored, and at vanishing bias a current is generated (*dashed, dotted,* and *dash–dotted curves*). At infinite feedback strength, transport becomes unidirectional and the current is completely rectified (*solid orange*). When the current curves enter the top left quadrant (*dashed brown lines*), the power generated by the device $P = -IV$ becomes positive. Other parameters: $\varepsilon = 0$, tunneling rates parametrized as in Eq. (7.65), $\mu_L = +V/2 = -\mu_R$ (Color figure online)

Taking a bird's eye view, we have applied two Liouvillians that if applied alone would have led to transport with the bias. By applying them conditioned on the system state, we apply both Liouvillians in a turnstyle, where the durations however depend on the electronic dwell times. The fact that two losing strategies combined may sometimes lead to a winning strategy is also known as Parrondo's paradox [8].

There are some minor differences to the original proposal by Maxwell: first, the demon does not monitor molecular speeds but instead whether a charge is localized on the dot or not. Second, the demon has two shutters instead of one to control the transfer of electrons. Third, for finite feedback strength, the demon's control only works probabilistically; i.e., the demon is not perfect. Nevertheless, the net effect is quite similar: starting at equilibrium $V = 0$ and leaving the bias voltage variable, the feedback control loop will generate an increasing bias until its finite feedback strength will not be able to increase it further and a new stationary state with a vanishing current at finite bias is reached [9]. Below, we will discuss how the entropy balance is modified by the feedback control loop.

7.3.3 Entropy Production

The discussed case however has an interesting thermodynamic interpretation: in the entropy flow in Def. (4.3), the modification of the local detailed balance condition (7.62) will lead to an additional contribution

$$\dot{S}_e = \dot{S}_e^{\text{app}} + \mathscr{I}, \tag{7.68}$$

where $\dot{S}_e^{\text{app}} = \beta_L(I_E^{(L)} - \mu_L I_M^{(L)}) + \beta_R(I_E^{(R)} - \mu_L I_M^{(R)})$ is the apparent entropy flow that can be reconstructed from the heat flows to the left and right reservoirs, and the additional contribution is the information current

$$\mathscr{I} = \sum_{ij\in\{E,F\}} \sum_{v\in\{L,R\}} L_{ij}^{(v)} \bar{P}_j \delta_{ij}^{(v)}. \tag{7.69}$$

We note that the above defined information current should not be confused with the mutual information [10–13] between system and demon dots: most obvious, it can become negative, whereas the mutual information is positive definite. Since at the modified steady state, entropy production balances the entropy flow, we also modify the second law of thermodynamics. Including the information current, the entropy production is still positive:

$$\dot{S}_i = -\dot{S}_e = -\dot{S}_e^{\text{app}} - \mathscr{I} = +\dot{S}_i^{\text{app}} - \mathscr{I} \geq 0. \tag{7.70}$$

We may however now well generate situations where the apparent entropy production becomes negative $\dot{S}_i^{\text{app}} < 0$. Considering as an example equal temperatures $\beta_L = \beta_R = \beta$, the average apparent entropy production and information current become

$$\dot{S}_i^{\text{app}} = I_M^L \beta(\mu_L - \mu_R), \qquad \mathscr{I} = I_M^L \ln\frac{\Gamma_L^E \Gamma_R^F}{\Gamma_L^F \Gamma_R^E} \overset{\text{Eq. (7.65)}}{\rightarrow} 4\delta I_M^L. \tag{7.71}$$

Inserting the apparent entropy production into the total one, we identify in the interesting region a positive power output as a negative contribution

$$\dot{S}_i = -\beta P - \mathscr{I} \geq 0, \tag{7.72}$$

which must be over-balanced by the information current. With this one can define the efficiency of information-to-power conversion in the range $-4\delta/\beta \leq V \leq 0$ (which corresponds to positive power output),

$$\eta = \frac{\beta P}{-\mathscr{I}} = -\frac{\beta V}{4\delta} \leq 1. \tag{7.73}$$

At vanishing voltage, the generated feedback current is largest, but the power vanishes, and the efficiency is zero. Likewise, at vanishing current, where $V = V^* =$

$-4\delta/\beta$, the efficiency reaches one, but the device is useless since the output power vanishes. Therefore, it is customary to consider the efficiency at maximum power

$$\bar{\eta} = -\frac{\beta \bar{V}}{4\delta}, \tag{7.74}$$

where we see that it vanishes at infinite feedback strength $\delta \to \infty$. For finite feedback strength however, one can first obtain the voltage at which power output is maximal and then relate this with the required feedback strength. For the dashed ($\delta = 1/2$), dotted ($\delta = 1$), and dash–dotted ($\delta = 2$) curves in Fig. 7.10 we obtain the efficiencies $\bar{\eta} = 0.48$, $\bar{\eta} = 0.43$, and $\bar{\eta} = 0.30$, respectively.

7.4 Self-Controlling Systems: A Complete Description of Maxwell's Demon

In any case, the discussed feedback control systems require an extremely fast processing of the measurement signal and a similarly fast control action. For the discussed master equation to remain valid, these time scales must be much shorter than the decay time of the open quantum systems. Whereas for electrons trapped in quantum dots without coherences the dwell times may reach hours, such that signal processing and feedback control may be processed much faster, this may appear questionable for more delicate quantum systems. Although the existent approaches can be extended by introducing a delay between measurement and control [14, 15], it would be interesting to learn which physical systems by their intrinsic evolution yield—when a subsystem is considered—the same effective evolution as if the subsystem were subject to an infinitely fast classical feedback control loop. Such systems would conveniently control themselves and minimize the delay. Fortunately, such a system exists for the Maxwell demon discussed in Sect. 7.3. In the following, we will discuss the example provided in Ref. [16].

Consider a single electron transistor (SET) now capacitively interacting with another quantum dot that is coupled to its own reservoir, as depicted in Fig. 7.11. The system Hamiltonian of this three-terminal system reads as

$$H_{\mathrm{S}} = \varepsilon_d c_d^\dagger c_d + \varepsilon_s c_s^\dagger c_s + U c_d^\dagger c_d c_s^\dagger c_s, \tag{7.75}$$

where ε_s and ε_d denote the on-site energies of the SET dot and the demon dot, respectively, whereas U denotes the Coulomb interaction between the two dots. The system dot is tunnel-coupled to the left and right leads, whereas the demon dot is tunnel-coupled to its junction only:

$$H_{\mathrm{I}} = \sum_k \left(t_{kL} c_s c_{kL}^\dagger + t_{kL}^* c_{kL} c_s^\dagger \right) + \sum_k \left(t_{kR} c_s c_{kR}^\dagger + t_{kR}^* c_{kR} c_s^\dagger \right)$$
$$+ \sum_k \left(t_{kd} c_d c_{kd}^\dagger + t_{kd}^* c_{kL} c_d^\dagger \right). \tag{7.76}$$

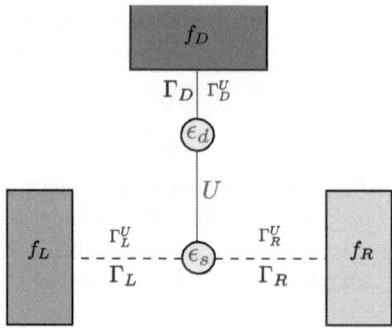

Fig. 7.11 Sketch of an SET (*bottom circuit*) that is capacitively coupled via the Coulomb interaction U to another quantum dot. The additional quantum dot is tunnel-coupled to its own reservoir with Fermi function f_D. Since the associated stationary matter current vanishes, only energy is continuously transferred across this junction (*dotted line*). The energy-dependent tunneling rates $\Gamma_\alpha(\omega)$ are evaluated at the system's transition frequencies, which leads to multiple tunneling rates for each junction such as, e.g., $\Gamma_D^U = \Gamma_D(\varepsilon_d + U)$ and $\Gamma_D = \Gamma_D(\varepsilon)$. Neither open-loop nor feedback control is at work, but the dot of the SET—if considered separately—behaves in certain parameter regions as if subjected to an external feedback control loop

Furthermore, all the junctions are modeled as noninteracting fermions:

$$H_{\mathrm{B}} = \sum_{\nu \in \{L,R,d\}} \sum_k \varepsilon_{k\nu} c_{k\nu}^\dagger c_{k\nu}. \tag{7.77}$$

7.4.1 Derivation of the Rate Equation

Treating the tunneling amplitudes perturbatively and fixing the reservoirs at thermal equilibrium states

$$\bar{\rho}_{\mathrm{B}}^{(\nu)} = \frac{e^{-\beta_\nu (H_{\mathrm{B}}^{(\nu)} - \mu_\nu N_{\mathrm{B}}^{(\nu)})}}{\mathrm{Tr}\{e^{-\beta_\nu (H_{\mathrm{B}}^{(\nu)} - \mu_\nu N_{\mathrm{B}}^{(\nu)})}\}}, \tag{7.78}$$

we use

$$\mathrm{Tr}\{c_{k\nu}^\dagger c_{k'\nu} \bar{\rho}_{\mathrm{B}}^{(\nu)}\} = \delta_{kk'} f_\nu(\varepsilon_{k\nu}),$$
$$\mathrm{Tr}\{c_{k\nu} c_{k'\nu}^\dagger \bar{\rho}_{\mathrm{B}}^{(\nu)}\} = \delta_{kk'} [1 - f_\nu(\varepsilon_{k\nu})] \tag{7.79}$$

to derive the standard quantum optical master equation (compare also Definition 2.3). Furthermore, we assume a continuous spectrum of reservoir frequencies by assuming that the tunneling rates

$$\Gamma_\nu(\omega) = 2\pi \sum_k |t_{k\nu}|^2 \delta(\omega - \varepsilon_{k\nu}) \tag{7.80}$$

are a continuous function of ω. Importantly, we do not apply the popular wide-band limit here (which would mean approximating $\Gamma_\nu(\omega) \approx \Gamma_\nu$).

In the energy eigenbasis of H_S—further on denoted by $|\rho\sigma\rangle$, where $\rho \in \{E, F\}$ describes the system's dot state and $\sigma \in \{0, 1\}$ denotes the state of the demon dot (both either empty or filled, respectively)—the populations obey a simple rate equation defined by Eq. (2.47). Denoting the populations by $p_{\rho\sigma} = \langle\rho\sigma|\rho|\rho\sigma\rangle$, the rate equation $\dot{P} = \mathscr{L}P$ in the ordered basis $P = (p_{0E}, p_{1E}, p_{0F}, p_{1F})^T$ decomposes into the contributions due to the different reservoirs $\mathscr{L} = \mathscr{L}_D + \mathscr{L}_L + \mathscr{L}_R$, which read

$$\mathscr{L}_D = \begin{pmatrix} -\Gamma_D f_D & +\Gamma_D(1 - f_D) & 0 & 0 \\ +\Gamma_D f_D & -\Gamma_D(1 - f_D) & 0 & 0 \\ 0 & 0 & -\Gamma_D^U f_D^U & +\Gamma_D^U(1 - f_D^U) \\ 0 & 0 & +\Gamma_D^U f_D^U & -\Gamma_D^U(1 - f_D^U) \end{pmatrix},$$

$$\mathscr{L}_\alpha = \begin{pmatrix} -\Gamma_\alpha f_\alpha & 0 & +\Gamma_\alpha(1 - f_\alpha) & 0 \\ 0 & -\Gamma_\alpha^U f_\alpha^U & 0 & +\Gamma_\alpha^U(1 - f_\alpha^U) \\ +\Gamma_\alpha f_\alpha & 0 & -\Gamma_\alpha(1 - f_\alpha) & 0 \\ 0 & +\Gamma_\alpha^U f_\alpha^U & 0 & -\Gamma_\alpha^U(1 - f_\alpha^U) \end{pmatrix}, \quad \alpha \in \{L, R\},$$

(7.81)

where we have used the abbreviations $\Gamma_\alpha = \Gamma_\alpha(\varepsilon_s)$ and $\Gamma_\alpha^U = \Gamma_\alpha(\varepsilon_s + U)$ for $\alpha \in \{L, R\}$ and $\Gamma_D = \Gamma_D(\varepsilon_d)$ and $\Gamma_D^U = \Gamma_D(\varepsilon_d + U)$ for the tunneling rates and similarly for the Fermi functions $f_\alpha = f_\alpha(\varepsilon_s)$, $f_\alpha^U = f_\alpha(\varepsilon_s + U)$, $f_D = f_D(\varepsilon_d)$, and $f_D^U = f_D(\varepsilon_d + U)$, respectively. We note that all contributions separately obey local detailed balance relations (4.43). We also note that the above rate matrix constitutes a special case of the model in Sect. 5.3, when one junction is disconnected. Closer inspection of the rates in Eq. (7.81) reveals that these rates could have been guessed without any microscopic derivation. For example, the transition rate from state $|1E\rangle$ to state $|0E\rangle$ is just given by the bare tunneling rate for the demon junction Γ_D multiplied by the probability of finding a free space in the terminal at transition frequency ε_d. Similarly, the transition rate from state $|1F\rangle$ to state $|0F\rangle$ corresponds to an electron jumping out of the demon dot to its junction, but this time transporting energy of $\varepsilon_d + U$. We have ordered our basis such that the upper left block of $\mathscr{L}_D$ describes the dynamics of the demon dot conditioned on an empty system dot, whereas the lower block accounts for the dynamics conditioned on a filled system dot.

7.4.2 Counting Statistics and Entropy

As a whole, the system respects the second law of thermodynamics. We demonstrate this by analyzing the entropy production by means of the full counting statistics. In order to avoid having to trace six counting fields, we note that the system obeys three conservation laws, since the two dots may only exchange energy but not matter:

$$I_M^{(L)} + I_M^{(R)} = 0, \qquad I_M^{(D)} = 0, \qquad I_E^{(L)} + I_E^{(R)} + I_E^{(D)} = 0, \qquad (7.82)$$

where $I_E^{(\nu)}$ and $I_M^{(\nu)}$ denote energy and matter currents to terminal ν, respectively. Therefore, three counting fields should in general suffice to completely track the entropy production in the long-term limit. For simplicity however, we compute the entropy production for the more realistic case of equal temperatures at the left and right SET junctions $\beta = \beta_L = \beta_R$. Technically, this is conveniently performed by balancing with the entropy flow and using the conservation laws

$$
\begin{aligned}
\dot{S}_i = -\dot{S}_e &= -\sum_\nu \beta^{(\nu)}\big(I_E^{(\nu)} - \mu^{(\nu)} I_M^{(\nu)}\big) \\
&= -\beta\big(I_E^{(L)} - \mu_L I_M^{(L)} + I_E^{(R)} - \mu_R I_M^{(R)}\big) - \beta_D I_E^{(D)} \\
&= (\beta - \beta_D) I_E^{(D)} - \beta(\mu_L - \mu_R) I_M^{(R)}.
\end{aligned}
\qquad (7.83)
$$

Thus, we conclude that for equal temperatures left and right it should even suffice to track, e.g., only the energy transferred to the demon junction and the particles to the right lead. Therefore, we introduce counting fields for the demon (ξ) and for the particles transferred to the left junctions (χ), and the counting-field-dependent rate equation becomes

$$\mathscr{L}_D(\xi)$$

$$
= \begin{pmatrix}
-\Gamma_D f_D & +\Gamma_D(1-f_D)e^{+i\xi\varepsilon_d} & 0 & 0 \\
+\Gamma_D f_D e^{-i\xi\varepsilon_d} & -\Gamma_D(1-f_D) & 0 & 0 \\
0 & 0 & -\Gamma_D^U f_D^U & +\Gamma_D^U(1-f_D^U)e^{+i\xi(\varepsilon_d+U)} \\
0 & 0 & +\Gamma_D^U f_D^U e^{-i\xi(\varepsilon_d+U)} & -\Gamma_D^U(1-f_D^U)
\end{pmatrix},
$$

$$
\mathscr{L}_R(\chi) = \begin{pmatrix}
-\Gamma_R f_R & 0 & +\Gamma_R(1-f_R)e^{+i\chi} & 0 \\
0 & -\Gamma_R^U f_R^U & 0 & +\Gamma_R^U(1-f_R^U)e^{+i\chi} \\
+\Gamma_R f_R e^{-i\chi} & 0 & -\Gamma_R(1-f_R) & 0 \\
0 & +\Gamma_R^U f_R^U e^{-i\chi} & 0 & -\Gamma_R^U(1-f_R^U)
\end{pmatrix}.
$$

$$(7.84)$$

These counting fields can now be used to reconstruct the statistics of energy and matter transfer. The currents can be obtained by performing suitable derivatives of the rate matrix; compare Eq. (4.76). For example, the energy current to the demon is given by $I_E^{(D)} = -i\,\mathrm{Tr}\{\partial_\xi \mathscr{L}(\xi,0)|_{\xi=0}\bar{\rho}\}$, where $\bar{\rho}$ is the steady state $\mathscr{L}(0,0)\bar{\rho} = 0$.

To test the fluctuation theorem, we calculate the characteristic polynomial

$$
\begin{aligned}
\mathscr{D}(\xi,\chi) &= \big|\mathscr{L}(\xi,\chi) - \lambda\mathbf{1}\big| \\
&= (L_{11} - \lambda)(L_{22} - \lambda)(L_{33} - \lambda)(L_{44} - \lambda) \\
&\quad - (L_{11} - \lambda)(L_{22} - \lambda)L_{34}(\xi)L_{43}(\xi) \\
&\quad - (L_{11} - \lambda)(L_{33} - \lambda)L_{24}(\chi)L_{42}(\chi)
\end{aligned}
$$

$$- (L_{22} - \lambda)(L_{44} - \lambda)L_{13}(\chi)L_{31}(\chi)$$

$$- (L_{33} - \lambda)(L_{44} - \lambda)L_{12}(\xi)L_{21}(\xi)$$

$$+ L_{12}(\xi)L_{21}(\xi)L_{34}(\xi)L_{43}(\xi) + L_{13}(\chi)L_{31}(\chi)L_{24}(\chi)L_{42}(\chi)$$

$$- L_{12}(\xi)L_{24}(\chi)L_{31}(\chi)L_{43}(\xi) - L_{13}(\chi)L_{21}(\xi)L_{34}(\xi)L_{42}(\chi)$$

$$= (L_{11} - \lambda)(L_{22} - \lambda)(L_{33} - \lambda)(L_{44} - \lambda)$$

$$- (L_{11} - \lambda)(L_{22} - \lambda)L_{34}(0)L_{43}(0)$$

$$- (L_{11} - \lambda)(L_{33} - \lambda)L_{24}(\chi)L_{42}(\chi)$$

$$- (L_{22} - \lambda)(L_{44} - \lambda)L_{13}(\chi)L_{31}(\chi)$$

$$- (L_{33} - \lambda)(L_{44} - \lambda)L_{12}(0)L_{21}(0)$$

$$+ L_{12}(0)L_{21}(0)L_{34}(0)L_{43}(0) + L_{13}(\chi)L_{31}(\chi)L_{24}(\chi)L_{42}(\chi)$$

$$- L_{12}(\xi)L_{24}(\chi)L_{31}(\chi)L_{43}(\xi) - L_{13}(\chi)L_{21}(\xi)L_{34}(\xi)L_{42}(\chi), \quad (7.85)$$

where L_{ij} simply denote the matrix elements of the rate matrix $\mathscr{L}$. We note the symmetries

$$L_{13}(-\chi) = \frac{1 - f_L}{f_L} L_{31}\left(+\chi + i\ln\frac{f_L(1 - f_R)}{(1 - f_L)f_R}\right)$$

$$= \frac{1 - f_L}{f_L} L_{31}\left(+\chi + i\beta(\mu_L - \mu_R)\right),$$

$$L_{24}(-\chi) = \frac{1 - f_L^U}{f_L^U} L_{42}\left(+\chi + i\ln\frac{f_L^U(1 - f_R^U)}{(1 - f_L^U)f_R^U}\right)$$

$$= \frac{1 - f_L^U}{f_L^U} L_{42}\left(+\chi + i\beta(\mu_L - \mu_R)\right), \qquad (7.86)$$

$$L_{12}(-\xi) = L_{21}\left(+\xi + \frac{i}{\varepsilon_d}\ln\frac{1 - f_D}{f_D}\right) = L_{21}\left(+\xi + \frac{i}{\varepsilon_d}\beta_D(\varepsilon_d - \mu_D)\right),$$

$$L_{34}(-\xi) = L_{43}\left(+\xi + \frac{i}{\varepsilon_d + U}\ln\frac{1 - f_D^U}{f_D^U}\right)$$

$$= L_{43}\left(+\xi + \frac{i}{\varepsilon_d + U}\beta_D(\varepsilon_d + U - \mu_D)\right),$$

which can be used to show that the full characteristic polynomial obeys the symmetry

$$\mathscr{D}(-\xi, -\chi) = \mathscr{D}\left(\xi + i(\beta_D - \beta)/U, \chi + i\beta(\mu_L - \mu_R)\right). \qquad (7.87)$$

This symmetry implies—when monitoring the energy current to the demon e_D and the number of electrons transferred to the right junction n_R—for the corresponding probability distribution the fluctuation theorem

$$\lim_{t \to \infty} \frac{P_{+\Delta n_S, +\Delta e_D}}{P_{-\Delta n_S, -\Delta e_D}} = e^{(\beta_D - \beta)\Delta e_D + \beta(\mu_L - \mu_R)\Delta n_S}. \tag{7.88}$$

Instead of determining the continuous energy emission distribution, we could alternatively have counted the discrete number of electrons entering the demon dot at energy ε_D and leaving it at energy $\varepsilon_D + U$. Since this process leads to a net energy extraction of energy U from the system, the corresponding matter current is tightly coupled to the energy current across the demon junction; i.e., their number would be related to the energy via $\Delta e_D = n_D U$. Comparing the value in the exponent of Eq. (7.88) with the average expectation value of the entropy production in Eq. (7.83), we can also roughly interpret the fluctuation theorem as the ratio of probabilities for trajectories with a positive and negative entropy production. Such findings can be made much more concrete for finite times by taking initial and final internal system configurations into account [17].

7.4.3 Global View: A Thermoelectric Device

From the discussion in Sect. 4.3 one concludes that the average value of Eq. (7.83) must be positive. In addition, we identify $P = (\mu_L - \mu_R)I_M^{(R)} = -(\mu_L - \mu_R)I_M^{(L)}$ as the power generated by the device, which—when the current flows against the bias—may yield a negative contribution βP to the overall entropy production. In these parameter regimes however, the negative contribution $\beta(\mu_L - \mu_R)I_M^{(R)}$ must be over-balanced by the second term $(\beta - \beta_D)I_E^{(D)}$, which clearly requires—when the demon reservoir is colder than the SET reservoirs $\beta_D > \beta_S$—the energy current to flow out of the demon, $I_E^{(D)} < 0$. As a whole, the system therefore just converts a thermal gradient between the two subsystems into power: a fraction of the heat coming from the hot SET leads is converted into power, and the remaining fraction is dissipated as heat at the cold demon junction. The corresponding efficiency for this conversion can be constructed from the output power $P = -(\mu_L - \mu_R)I_M^{(L)}$ and the input heat $\dot{Q}_L + \dot{Q}_R = -I_E^{(D)} - (\mu_L - \mu_R)I_M^{(L)} = \dot{Q}_{\text{diss}} + P$, where $\dot{Q}_{\text{diss}} = -I_E^{(D)}$ is the heat dissipated into the demon reservoir. Using $\dot{S}_i \geq 0$ we find that the efficiency—which of course is only useful in parameter regimes where the power is positive $\beta(\mu_L - \mu_R)I_M^{(R)} > 0$—is upper bounded by the Carnot efficiency:

$$\eta = \frac{P}{\dot{Q}_{\text{diss}} + P} \leq 1 - \frac{T_D}{T} = \eta_{\text{Carnot}}. \tag{7.89}$$

For practical applications a large efficiency is not always sufficient. For example, a maximum efficiency at zero power output would be quite useless. Therefore, it has become a common standard to first maximize the power output of the device and then compute the corresponding efficiency at maximum power. Due to the non-linearity of the underlying equations, this may be a difficult numerical optimization problem. To reduce the number of parameters, we assume that $f_D^U = 1 - f_D$ (which is the case when $\varepsilon_D = \mu_D - U/2$) and $f_L^U = 1 - f_R$ as well as $f_R^U = 1 - f_L$ (which for $\beta_L = \beta_R = \beta$ is satisfied when $\varepsilon_S = 1/2(\mu_L + \mu_R) - U/2$). See also the left panel of Fig. 7.12. Furthermore, we parametrize the modification of the tunneling rates by a single parameter via

$$\Gamma_L = \Gamma \frac{e^{+\delta}}{\cosh(\delta)}, \qquad \Gamma_L^U = \Gamma \frac{e^{-\delta}}{\cosh(\delta)},$$
$$\Gamma_R = \Gamma \frac{e^{-\delta}}{\cosh(\delta)}, \qquad \Gamma_R^U = \Gamma \frac{e^{+\delta}}{\cosh(\delta)} \tag{7.90}$$

to favor transport in a particular direction. We have inserted the normalization by $\cosh(\delta)$ to keep the tunneling rates finite as the feedback strength δ is increased. Trivially, at $\delta = 0$ we recover symmetric unperturbed tunneling rates and when $\delta \to \infty$, transport will be completely rectified. The matter current from left to right in the limit where the demon dot is much faster than the SET ($\Gamma_D \to \infty$ and $\Gamma_D^U \to \infty$) becomes

$$I_M^{(L)} = \frac{\Gamma}{2} \big[f_L - f_R + \tanh(\delta)(f_L + f_R - 2f_D) \big]. \tag{7.91}$$

Similarly, we obtain for the energy current to the demon

$$I_E^{(D)} = \frac{\Gamma U}{2} \big[f_L + f_R - 2f_D + (f_L - f_R)\tanh(\delta) \big], \tag{7.92}$$

which determines the dissipated heat. These can be converted into an efficiency solely expressed by Fermi functions when we use

$$\beta(\mu_L - \mu_R) = \ln\left(\frac{f_L(1 - f_R)}{(1 - f_L)f_R} \right),$$
$$\beta U = \ln\left(\frac{f_R(1 - f_R^U)}{(1 - f_R)f_R^U} \right) \to \ln\left(\frac{f_R f_L}{(1 - f_R)(1 - f_L)} \right), \tag{7.93}$$

which can be used to write the efficiency of heat to power conversion as

$$\eta = \frac{P}{\dot{Q}_{\text{diss}} + P} = \frac{1}{1 + \frac{\beta \dot{Q}_{\text{diss}}}{\beta P}}$$
$$= \frac{1}{1 + \frac{\ln(\frac{f_R f_L}{(1 - f_R)(1 - f_L)})(f_L + f_R - 2f_D + (f_L - f_R)\tanh(\delta))}{\ln(\frac{f_L(1 - f_R)}{(1 - f_L)f_R})(f_L - f_R + (f_L + f_R - 2f_D)\tanh(\delta))}}, \tag{7.94}$$

which is also illustrated in Fig. 7.12.

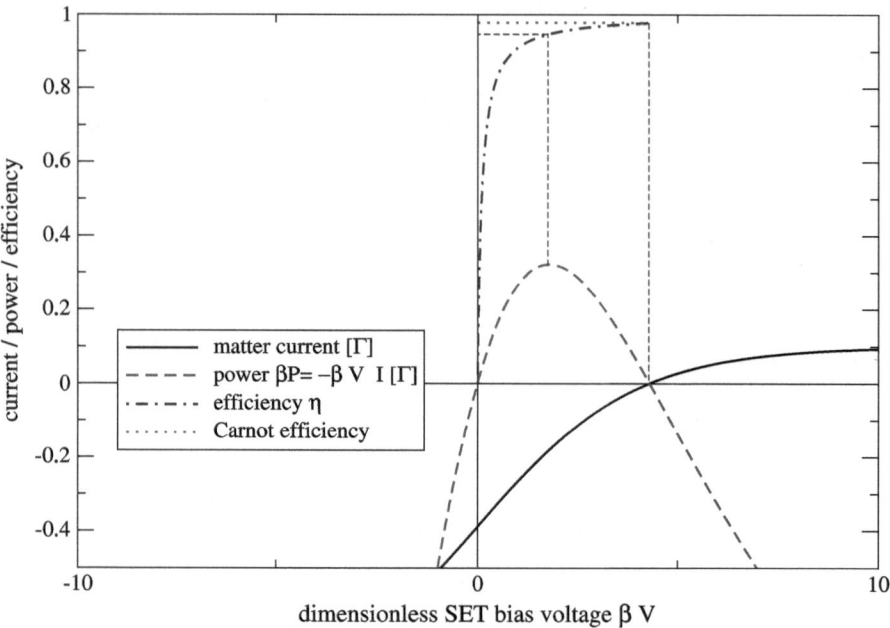

Fig. 7.12 *Top*: Sketch of the assumed configurations of chemical potentials, which imply at $\beta_L = \beta_R$ relations between the Fermi functions. *Bottom*: Plot of current (*solid black*, in units of Γ), dimensionless power $\beta V I$ (*dashed red*, in units of Γ), and efficiency η (*dash–dotted blue*) versus dimensionless bias voltage. At equilibrated bias (*origin*), the efficiency vanishes by construction, whereas it reaches Carnot efficiency (*dotted green*) at the new equilibrium, i.e., at zero power. At maximum power however, the efficiency still closely approaches the Carnot efficiency. Parameters: $\delta = 100$, tunneling rates parametrized as in Eq. (7.90), $f_D = 0.9 = 1 - f_D^U$, $\beta\varepsilon_S = -0.05 = -\beta(\varepsilon_S + U)$, such that the Carnot efficiency becomes $\eta_{\text{Carnot}} = 1 - (\beta U)/(\beta_D U) \approx 0.977244$

Beyond these average considerations, the qualitative action of the device may also be understood at the level of single trajectories; see Fig. 7.13. It should be noted that, at the trajectory level, all possible trajectories are still allowed, even

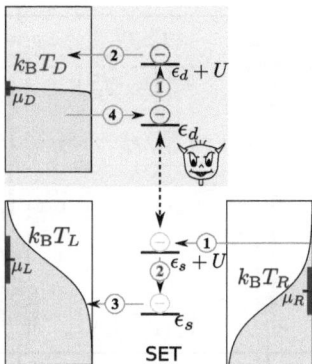

Fig. 7.13 Level sketch of the setup, adapted from Ref. [16]. *Shaded regions* represent occupied levels in the leads with chemical potentials and temperatures indicated. *Central horizontal lines* represent transition energies of system and demon dot, respectively. When the other dot is occupied, the bare transition frequency of every system is shifted by the Coulomb interaction U. The shown trajectory then becomes likely in the suggested Maxwell-demon mode: initially, the SET is empty and the demon dot is filled. When $\Gamma_R^U \gg \Gamma_L^U$, the SET dot is most likely first filled from the left lead, which shifts the transition frequency of the demon (1). When the bare tunneling rates of the demon are much larger than those of the SET, the demon dot will rapidly equilibrate by expelling the electron to its associated reservoir (2) before a further electronic jump at the SET can occur. At the new transition frequency, the SET electron is more likely to escape first to the left than to the right when $\Gamma_L \gg \Gamma_R$ (3). Now, the demon dot will equilibrate again by filling with an electron (4), thus restoring the initial state. In essence, an electron is transferred against the bias through the SET circuit while in the demon system an electron enters at energy ε_d and leaves at energy $\varepsilon_d + U$, leading to a net transfer of U from the demon into its reservoir

though those with positive total entropy production must on average dominate. As a whole, the system thereby merely converts a temperature gradient (cold demon, hot system) into useful power (current times voltage). A similar interpretation as a thermoelectric generator is possible when the temperature gradient is reversed, i.e., when the demon junction is hot and the SET junctions are cold [18].

7.4.4 Local View: A Feedback-Controlled Device

An experimentalist having access only to the SET circuit would measure a positive generated power and conserved particle currents $I_M^{(L)} + I_M^{(R)} = 0$, but possibly a slight mismatch of left and right energy currents $I_E^{(L)} + I_E^{(R)} = -I_E^{(D)} \neq 0$. This mismatch could not fully account for the generated power, since for any efficiency $\eta > 1/2$ in Fig. 7.13 we have $|I_E^{(D)}| < P$. Therefore, the experimentalist would conclude that his description of the system by energy and matter flows is not complete and he might suspect Maxwell's demon at work. Here, we will make the reduced dynamics of the SET dot alone more explicit by deriving a reduced rate equation.

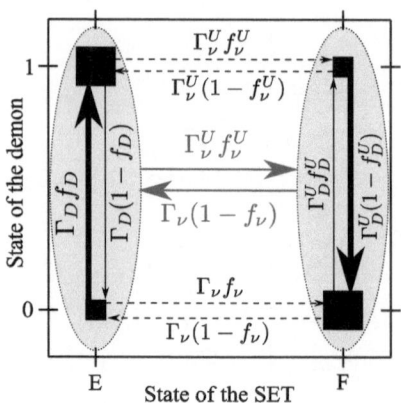

Fig. 7.14 Sketch of the coarse-graining procedure, taken from Ref. [16]. *Squares* represent states of the combined system, where *thicknesses of arrows* and *squares* denote relative stationary occupations and relative value of transition rates, respectively. When the demon transitions (*solid arrows*) are much faster than the SET transitions (*dashed arrows*), it is possible to obtain a coarse-grained dynamics between mesostates (*shaded oval areas*) that account either for an empty or a filled SET dot. The transition rates depicted in the center account for the error-free limit where $f_D \to 1$ and $f_D^U \to 0$

We can evidently write the rate equation defined by Eqs. (7.81) as $\dot{P}_{\rho\sigma} = \mathscr{L}_{\rho\sigma,\rho'\sigma'} P_{\rho'\sigma'}$, where ρ and σ label the demon and system degrees of freedom, respectively. If we discard the dynamics of the demon dot by tracing over its degrees of freedom $P_\sigma = \sum_\rho P_{\rho\sigma}$, we formally arrive at a non-Markovian evolution equation for the populations of the SET dot:

$$\dot{P}_\sigma = \sum_{\sigma'} \sum_{\rho\rho'} \mathscr{L}_{\rho\sigma,\rho'\sigma'} P_{\rho'\sigma'} = \sum_{\sigma'} \left[\sum_{\rho\rho'} \mathscr{L}_{\rho\sigma,\rho'\sigma'} \frac{P_{\rho'\sigma'}}{P_{\sigma'}} \right] P_{\sigma'}. \qquad (7.95)$$

Here, we may identify $\frac{P_{\rho'\sigma'}}{P_{\sigma'}}$ as the conditional probability of the demon being in state ρ' provided the system is in state σ'.

However, direct inspection of the rates suggests that when we assume the limit where the bare rates of the demon system are much larger than the SET tunneling rates, these conditional probabilities will assume their conditioned stationary values much faster than the SET dynamics. In this limit, the dynamics is mainly dominated by transitions between just two mesostates instead of the original four states; see Fig. 7.14. The mesostates are associated to either a filled or an empty system quantum dot, respectively. We may hence arrive again at a Markovian description by approximating

$$\frac{P_{\rho'\sigma'}}{P_{\sigma'}} \to \frac{\bar{P}_{\rho'\sigma'}}{\bar{P}_{\sigma'}}, \qquad (7.96)$$

which yields the coarse-grained rate matrix

$$\mathcal{W}_{\sigma\sigma'} = \sum_{\rho\rho'} \mathcal{L}_{\rho\sigma,\rho'\sigma'} \frac{\bar{P}_{\rho'\sigma'}}{\bar{P}_{\sigma'}}. \tag{7.97}$$

For the model at hand, the stationary conditional probabilities become in the limit where $\Gamma_D^{(U)} \gg \Gamma_{L/R}^{(U)}$

$$P_{0|E} = \frac{\bar{P}_{0E}}{\bar{P}_E} = 1 - f_D, \qquad P_{1|E} = \frac{\bar{P}_{1E}}{\bar{P}_E} = f_D,$$

$$P_{0|F} = \frac{\bar{P}_{0F}}{\bar{P}_F} = 1 - f_D^U, \qquad P_{1|F} = \frac{\bar{P}_{1F}}{\bar{P}_F} = f_D^U, \tag{7.98}$$

and just describe the fact that, due to the time-scale separation, the demon dot immediately reaches a thermal stationary state that depends on the occupation of the SET dot. The temperature and chemical potential of the demon reservoir determine if and how well the demon dot—which can be envisaged as the demon's memory capable of storing just one bit—captures the actual state of the system dot. For example, for high demon temperatures it will be roughly independent of the system dots occupation as $f_D \approx f_D^U \approx 1/2$. At very low demon temperatures however, and if the chemical potential of the demon dot is adjusted such that $\varepsilon_d - \mu_D < 0$ and $\varepsilon_d + U - \mu_D > 0$, the demon dot will nearly accurately (more formally when $\beta_D U \gg 1$) track the system occupation, since $f_D \to 1$ and $f_D^U \to 0$. Then, the demon dot will immediately fill when the SET dot is emptied and its electron will leave when the SET dot is filled. It thereby faithfully detects the state of the SET. In the presented model, the demon temperature thereby acts as a source of error in the demon's measurement of the system's state. In addition, the model at hand allows us to investigate the detector backaction on the probed system, which is often neglected. Here, this backaction is essential, and we will now investigate it by analyzing the reduced dynamics in detail.

The coarse-grained probabilities P_E and P_F of finding the SET dot empty or filled, respectively, obey the rate equation dynamics

$$\mathcal{L} = \begin{pmatrix} -L_{FE} & +L_{EF} \\ +L_{FE} & -L_{EF} \end{pmatrix} \tag{7.99}$$

with the coarse-grained rates

$$L_{EF} = L_{0E,0F} \frac{\bar{P}_{0F}}{\bar{P}_F} + L_{1E,1F} \frac{\bar{P}_{1F}}{\bar{P}_F}$$

$$= (1 - f_D^U)[\Gamma_L(1 - f_L) + \Gamma_R(1 - f_R)]$$

$$+ f_D^U[\Gamma_L^U(1 - f_L^U) + \Gamma_R^U(1 - f_R^U)], \tag{7.100}$$

$$L_{FE} = L_{0F,0E} \frac{\bar{P}_{0E}}{\bar{P}_E} + L_{1F,1E} \frac{\bar{P}_{1E}}{\bar{P}_E}$$

$$= (1 - f_D)[\Gamma_L f_L + \Gamma_R f_R] + f_D[\Gamma_L^U f_L^U + \Gamma_R^U f_R^U].$$

We note that a naive experimenter who is not aware of the demon interacting with the SET circuit would attribute the rates in the coarse-grained dynamics to just two reservoirs: $\mathcal{L} = \mathcal{L}_L + \mathcal{L}_R$ with the rates $\mathcal{L}_{EF}^{(\alpha)} = (1 - f_D^U)\Gamma_\alpha(1 - f_\alpha) + f_D^U \Gamma_\alpha^U(1 - f_\alpha^U)$ and $\mathcal{L}_{FE}^{(\alpha)} = (1 - f_D)\Gamma_\alpha f_\alpha + f_D \Gamma_\alpha^U f_\alpha^U$. Thus, when the SET is not sensitive to the demon state $\Gamma_{L/R}^U \approx \Gamma_{L/R}$ and $f_{L/R}^U \approx f_{L/R}$, local detailed balance is restored, and we recover the conventional SET rate equation (5.6).

We note that the matter current

$$I_M^{(v)} = L_{EF}^{(v)} \bar{P}_F - \mathcal{L}_{FE}^{(v)} \bar{P}_E \tag{7.101}$$

is conserved, $I_M^{(L)} = -I_M^{(R)}$, such that the entropy production in Definition (4.4) becomes

$$\dot{S}_i = \sum_{v \in \{L,R\}} L_{EF}^{(v)} \bar{P}_F \ln\left(\frac{\mathcal{L}_{EF}^{(v)} \bar{P}_F}{\mathcal{L}_{FE}^{(v)} \bar{P}_E}\right) + \mathcal{L}_{FE}^{(v)} \bar{P}_E \ln\left(\frac{\mathcal{L}_{FE}^{(v)} \bar{P}_E}{\mathcal{L}_{EF}^{(v)} \bar{P}_F}\right)$$

$$= \sum_{v \in \{L,R\}} (L_{EF}^{(v)} \bar{P}_F - \mathcal{L}_{FE}^{(v)} \bar{P}_E) \ln\left(\frac{\mathcal{L}_{EF}^{(v)} \bar{P}_F}{\mathcal{L}_{FE}^{(v)} \bar{P}_E}\right)$$

$$= I_M^{(L)} \ln\left(\frac{\mathcal{L}_{EF}^{(L)} \mathcal{L}_{FE}^{(R)}}{\mathcal{L}_{FE}^{(L)} \mathcal{L}_{EF}^{(R)}}\right) = I_M^{(L)} \mathscr{A}, \tag{7.102}$$

and is thus representable in a simple flux-affinity form. Similarly, we note that if we count particle transfers from the left to the right reservoir, the following fluctuation theorem would hold:

$$\frac{P_{+n}}{P_{-n}} = e^{n\mathscr{A}}, \tag{7.103}$$

and the fact that these fluctuations could in principle be resolved demonstrates that the affinity in the entropy production is a meaningful and measurable quantity. Without the demon dot, the conventional affinity of the SET would simply be given by

$$\mathscr{A}_0 = \ln\left(\frac{(1 - f_L)f_R}{f_L(1 - f_R)}\right) = \beta_L(\varepsilon - \mu_L) - \beta_R(\varepsilon - \mu_R), \tag{7.104}$$

and ignoring the physical implementation of the demon, we can interpret the modification of the entropy production due to the demon as an additional information current that is tightly coupled to the particle current:

$$\dot{S}_i = I_M^{(L)} \mathscr{A}_0 + I_M^{(L)}(\mathscr{A} - \mathscr{A}_0) = \dot{S}_i^{(0)} + \mathscr{I}. \tag{7.105}$$

When the demon temperature is lowered such that $\beta_D U \gg 1$ and its chemical potential is adjusted such that $f_D \to 1$ and $f_D^U \to 0$, the affinity becomes

$$\mathscr{A} = \ln\left(\frac{\Gamma_L(1-f_L)\Gamma_R^U f_R^U}{\Gamma_L^U f_L^U \Gamma_R(1-f_R)}\right) = \ln\left(\frac{\Gamma_L \Gamma_R^U}{\Gamma_L^U \Gamma_R}\right) + \ln\left(\frac{f_L f_R^U}{f_L^U f_R}\right) + \mathscr{A}_0. \quad (7.106)$$

The last term on the right-hand side is simply the affinity without the demon dot. The first two terms quantify the modification of the affinity. The pure limit of a Maxwell demon is reached, when the energetic backaction of the demon on the SET is negligible, i.e., when $f_L^U \approx f_L$ and $f_R^U \approx f_R$, which requires comparably large SET temperatures $\beta_{L/R} U \ll 1$. Of course, to obtain any nontrivial effect, it is still necessary to keep non-flat tunneling rates $\Gamma_{L/R}^U \neq \Gamma_{L/R}$, and in this case one recovers the case discussed in Sect. 7.3—identifying Γ_α^E with Γ_α and Γ_α^F with Γ_α^U.

7.5 Qubit Stabilization

The qubit, i.e., any quantum-mechanical two-level system that can be prepared in a superposition of its two states $|0\rangle$ and $|1\rangle$, is at the heart of quantum computers with great technological promise. One major obstacle to be overcome to build a quantum computer [19] is decoherence: qubits prepared in pure superposition states (as required for performing quantum computation) tend to decay into a statistical mixture when coupled to a destabilizing reservoir (of which there is an abundance in the real world). Here, we will approach the decoherence problem with a quantum master equation and we will show that feedback control can be used to act against the decay of coherences.

7.5.1 Model

The system is described by

$$H_S = \frac{\Omega}{2}\sigma^z, \qquad H_B^1 = \sum_k \omega_k^1 (b_k^1)^\dagger b_k^1, \qquad H_B^2 = \sum_k \omega_k^2 (b_k^2)^\dagger b_k^2$$

$$H_I^1 = \sigma^z \otimes \sum_k [h_k^1 b_k^1 + (h_k^1)^* (b_k^1)^\dagger], \qquad\qquad (7.107)$$

$$H_I^2 = \alpha(t)\sigma^x \otimes \sum_k [h_k^2 b_k^2 + (h_k^2)^* (b_k^2)^\dagger],$$

where σ^α represent the Pauli matrices and b_k the bosonic annihilation operators. Whereas the first interaction H_I^1 is of the pure dephasing type and would thus admit an exact solution (compare Sect. 3.1), the second is not, such that we will apply the master equation formalism for both interactions. We assume that the two bosonic baths are independent, such that we can calculate the dissipators separately. For clarity, we also assume that the two reservoirs are at the same temperature. The time dependence of $\alpha(t)$ is assumed to be piecewise constant and will be conditioned on a measurement result, which closes a feedback control loop. Since the reservoirs are at the same temperature, the bath correlation function is independent of the chosen interaction Hamiltonian (therefore we omit the indices):

$$
\begin{aligned}
C(\tau) &= \sum_{kk'} \left\langle \left(h_k b_k e^{-i\omega_k \tau} + h_k^* b_k^\dagger e^{+i\omega_k \tau} \right) \left(h_{k'} b_{k'} + h_{k'}^* b_{k'}^\dagger \right) \right\rangle \\
&= \sum_k |h_k|^2 \left[e^{-i\omega_k \tau} \left(1 + n(\omega_k) \right) + e^{+i\omega_k \tau} n(\omega_k) \right] \\
&= \frac{1}{2\pi} \int_0^\infty J(\omega) \left\{ e^{-i\omega\tau} \left[1 + n(\omega) \right] + e^{+i\omega\tau} n(\omega) \right\} d\omega, \qquad (7.108)
\end{aligned}
$$

where we have introduced the bosonic occupation $n(\omega) = [e^{\beta\omega} - 1]^{-1}$ with the inverse bath temperature β and the spectral density $J(\omega) = 2\pi \sum_k |h_k|^2 \delta(\omega - \omega_k)$. We would like to identify the Fourier transform of $C(\tau)$, such that we have to transform the integral to one involving the full real axis. Since all $\omega_k > 0$, the spectral density is defined for positive frequencies but can be analytically continued by defining $J(-\omega) = -J(+\omega)$. Making use of the relation $n(-\omega) = -[1 + n(\omega)]$, we rewrite the correlation function as

$$
\begin{aligned}
C(\tau) &= \frac{1}{2\pi} \int_0^\infty J(\omega) \left[1 + n(\omega) \right] e^{-i\omega\tau} \, d\omega + \frac{1}{2\pi} \int_{-\infty}^0 J(-\omega) n(-\omega) e^{-i\omega\tau} \, d\omega \\
&= \frac{1}{2\pi} \int_0^\infty J(\omega) \left[1 + n(\omega) \right] e^{-i\omega\tau} \, d\omega + \frac{1}{2\pi} \int_{-\infty}^0 J(\omega) \left[1 + n(\omega) \right] e^{-i\omega\tau} \, d\omega \\
&= \frac{1}{2\pi} \int_{-\infty}^\infty J(\omega) \left[1 + n(\omega) \right] e^{-i\omega\tau} \, d\omega. \qquad (7.109)
\end{aligned}
$$

We conclude for the Fourier transform of the bath correlation function,

$$
\gamma(\omega) = J(\omega) \left[1 + n(\omega) \right]. \qquad (7.110)
$$

It is easy to show that it fulfills the Kubo–Martin–Schwinger (KMS) condition

$$
\gamma(-\omega) = -J(\omega) \left[-n(\omega) \right] = J(\omega) n(\omega) = e^{-\beta\omega} \gamma(+\omega), \qquad (7.111)
$$

such that we may expect that the thermal state of the qubit with the reservoir temperature is one stationary state of the system. Note that due to the divergence of $n(\omega)$ at $\omega \to 0$, it is favorable to use an ohmic spectral density such as, e.g.,

$$
J(\omega) = J_0 \omega e^{-\omega/\omega_c}, \qquad (7.112)
$$

which grants an existing limit $\gamma(0)$. For a single system coupling operator A, the damping coefficients in the Born–Markov-secular approximation in Definition 2.3 simplify a bit:

$$\gamma_{ab,cd} = \delta(E_b - E_a, E_d - E_c)\gamma(E_b - E_a)\langle a|A|b\rangle\langle c|A|d\rangle^*$$

$$\sigma_{ab} = \delta_{E_a,E_b}\frac{1}{2i}\sum_c \sigma(E_b - E_c)\langle c|A|b\rangle\langle c|A|a\rangle^*, \tag{7.113}$$

where the odd Fourier transform $\sigma(\omega)$ of the bath correlation function can be obtained from $\gamma(\omega)$ by a Cauchy principal value integral (2.38) and the vectors denote the energy eigenbasis $\sigma^z|0\rangle = |0\rangle$ and $\sigma^z|1\rangle = -|1\rangle$. The Kronecker symbols automatically imply that many damping coefficients vanish, such that the action of the full Liouvillian can be written as

$$\dot{\rho} = -i\left[\left(\frac{\Omega}{2} + \sigma_{00}\right)|0\rangle\langle 0| + \left(-\frac{\Omega}{2} + \sigma_{11}\right)|1\rangle\langle 1|, \rho\right]$$

$$+ \gamma_{00,00}\left[|0\rangle\langle 0|\rho|0\rangle\langle 0| - \frac{1}{2}|0\rangle\langle 0|\rho - \frac{1}{2}\rho|0\rangle\langle 0|\right]$$

$$+ \gamma_{11,11}\left[|1\rangle\langle 1|\rho|1\rangle\langle 1| - \frac{1}{2}|1\rangle\langle 1|\rho - \frac{1}{2}\rho|1\rangle\langle 1|\right]$$

$$+ \gamma_{00,11}\left[|0\rangle\langle 0|\rho|1\rangle\langle 1|\right] + \gamma_{11,00}\left[|1\rangle\langle 1|\rho|0\rangle\langle 0|\right]$$

$$+ \gamma_{01,01}\left[|0\rangle\langle 1|\rho|1\rangle\langle 0| - \frac{1}{2}|1\rangle\langle 1|\rho - \frac{1}{2}\rho|1\rangle\langle 1|\right]$$

$$+ \gamma_{10,10}\left[|1\rangle\langle 0|\rho|0\rangle\langle 1| - \frac{1}{2}|0\rangle\langle 0|\rho - \frac{1}{2}\rho|0\rangle\langle 0|\right]. \tag{7.114}$$

It is a general feature of the Born–Markov-secular approximation that for a nondegenerate system Hamiltonian (here $\Omega \neq 0$) the populations and coherences in the energy eigenbasis decouple:

$$\dot{\rho}_{00} = -\gamma_{10,10}\rho_{00} + \gamma_{01,01}\rho_{11},$$

$$\dot{\rho}_{11} = +\gamma_{10,10}\rho_{00} - \gamma_{01,01}\rho_{11},$$

$$\dot{\rho}_{01} = \left[-\frac{1}{2}(\gamma_{00,00} + \gamma_{11,11} - 2\gamma_{00,11} + \gamma_{01,01} + \gamma_{10,10})\right.$$

$$\left. - i(\Omega + \sigma_{00} - \sigma_{11})\right]\rho_{01}, \tag{7.115}$$

$$\dot{\rho}_{10} = \left[-\frac{1}{2}(\gamma_{00,00} + \gamma_{11,11} - 2\gamma_{11,00} + \gamma_{01,01} + \gamma_{10,10})\right.$$

$$\left. + i(\Omega + \sigma_{00} - \sigma_{11})\right]\rho_{10}.$$

For the two interaction Hamiltonians chosen, we can make the corresponding coefficients explicit

Coefficient	A: pure dephasing $A = \sigma^z$	B: dissipation $A = \sigma^x$
$\gamma_{00,00}$	$+\gamma(0)$	0
$\gamma_{00,11}$	$-\gamma(0)$	0
$\gamma_{11,00}$	$-\gamma(0)$	0
$\gamma_{11,11}$	$+\gamma(0)$	0
$\gamma_{01,01}$	0	$\gamma(+\Omega)$
$\gamma_{10,10}$	0	$\gamma(-\Omega)$
σ_{00}	$\frac{\sigma(0)}{2i}$	$\frac{\sigma(-\Omega)}{2i}$
σ_{11}	$\frac{\sigma(0)}{2i}$	$\frac{\sigma(+\Omega)}{2i}$

and rewrite the corresponding Liouvillian in the ordering $\rho_{00}, \rho_{11}, \rho_{01}, \rho_{10}$ as a superoperator (further abbreviating $\gamma_{0/\pm} = \gamma(0/\pm\Omega)$, $\Sigma = \sigma_{00} - \sigma_{11}$)

$$
\mathcal{L}_A = \begin{pmatrix} 0 & 0 & 0 & 0 \\ 0 & 0 & 0 & 0 \\ 0 & 0 & -2\gamma_0 - i\Omega & 0 \\ 0 & 0 & 0 & -2\gamma_0 + i\Omega \end{pmatrix},
$$

$$
\mathcal{L}_B = \begin{pmatrix} -\gamma_- & +\gamma_+ & 0 & 0 \\ +\gamma_- & -\gamma_+ & 0 & 0 \\ 0 & 0 & -\frac{\gamma_-+\gamma_+}{2} - i(\Omega + \Sigma) & 0 \\ 0 & 0 & 0 & -\frac{\gamma_-+\gamma_+}{2} + i(\Omega + \Sigma) \end{pmatrix}.
$$

(7.116)

Keeping in mind that both Ω and Σ are real-valued, both Liouvillians therefore lead to a decay of coherences with rates (we assume $\Omega > 0$)

$$
\gamma_A = 2\gamma_0 = 2 \lim_{\omega \to 0} J(\omega)[1 + n(\omega)] = 2\frac{J_0}{\beta} = 2J_0 k_B T,
$$

$$
\gamma_B = \frac{\gamma_- + \gamma_+}{2} = \frac{1}{2}[J(\Omega)[1 + n(\Omega)] + J(-\Omega)[1 + n(-\Omega)]]
$$

$$
= \frac{1}{2}[J(\Omega)[1 + n(\Omega)] + J(\Omega)n(\Omega)]
$$

$$
= \frac{1}{2}J(\Omega) \coth\left[\frac{\Omega}{2k_B T}\right],
$$

(7.117)

which for large temperatures both scale proportionally to T. Therefore, in the high-temperature limit, the application of either Liouvillian or a superposition of both will simply lead to rapid decoherence.

The same can be expected from a turnstyle (open-loop control), where the Liouvillians act one at a time following a predefined protocol: due to the block structure of each Liouvillian, their exponential (forming the propagator) will also have the same block structure Consequently, the total propagator for any driving protocol will obey the same block structure, and the coherences will decay in the long-term limit.

7.5.2 Feedback Liouvillian

The situation changes however, when measurement results are used to determine which Liouvillian is acting. We choose to act with Liouvillian $\mathscr{L}_A$ throughout and to turn on Liouvillian $\mathscr{L}_B$ in addition—multiplied by a dimensionless feedback parameter $\alpha \geq 0$—when a certain measurement result is obtained. Given a measurement with just two outcomes, the effective propagator is then given by

$$\mathscr{W}(\Delta t) = e^{\mathscr{L}_A \Delta t} \mathscr{M}_1 + e^{(\mathscr{L}_A + \alpha \mathscr{L}_B)\Delta t} \mathscr{M}_2, \tag{7.118}$$

where $\mathscr{M}_i$ are the superoperators corresponding to the action of the measurement operators $M_i \rho M_i^\dagger$ on the density matrix. Any nontrivial effects can now only be expected when the block structure of the Liouvillians is lifted by the action of the measurement. Formally, the measurement superoperators should therefore not have the same block structure as the Liouvillians. Therefore, we consider a projective measurement of the σ^x expectation value (which does not commute with the system Hamiltonian). Formally, the corresponding measurement operators are given by

$$M_1 = \frac{1}{2}[\mathbf{1} + \sigma^x], \qquad M_2 = \frac{1}{2}[\mathbf{1} - \sigma^x]. \tag{7.119}$$

These projection operators obviously fulfill the completeness relation $M_1^\dagger M_1 + M_2^\dagger M_2 = \mathbf{1}$. The superoperators corresponding to $\mathscr{M}_i \rho \,\hat{=}\, M_i \rho M_i^\dagger$ are also orthogonal projectors:

$$\mathscr{M}_1 = \frac{1}{4}\begin{pmatrix} 1 & 1 & 1 & 1 \\ 1 & 1 & 1 & 1 \\ 1 & 1 & 1 & 1 \\ 1 & 1 & 1 & 1 \end{pmatrix}, \qquad \mathscr{M}_2 = \frac{1}{4}\begin{pmatrix} 1 & 1 & -1 & -1 \\ 1 & 1 & -1 & -1 \\ -1 & -1 & 1 & 1 \\ -1 & -1 & 1 & 1 \end{pmatrix}. \tag{7.120}$$

Exercise 7.8 (Measurement superoperators) Show the correspondence between M_i and $\mathscr{M}_i$ in the above equations.

In contrast to the feedback control scheme in the Maxwell demon implementation (compare Sect. 7.3), these measurement superoperators do not share the block structure of the Liouvillians. This implies that coherences can no longer be neglected, even in the long-term dynamics. In the model of Sect. 7.3 one would observe no effect if measurements were periodically applied without performing any control operations. Here, this is completely different: without feedback ($\alpha = 0$), the propagator becomes $\mathscr{P}(\Delta t) = e^{\mathscr{L}_A \Delta t}(\mathscr{M}_1 + \mathscr{M}_2)$, which is—since $\mathscr{M}_1 + \mathscr{M}_2 \neq 1$—already structurally different from the propagator without measurements and control $\mathscr{P}(\Delta t) = e^{\mathscr{L}_A \Delta t}$, as can be easily made explicit:

$$
e^{\mathscr{L}_A \Delta t}(\mathscr{P}_1 + \mathscr{P}_2) = \frac{1}{2}
\begin{pmatrix}
1 & 1 & 0 & 0 \\
1 & 1 & 0 & 0 \\
0 & 0 & e^{-(2\gamma_0+i\Omega)\Delta t} & e^{-(2\gamma_0+i\Omega)\Delta t} \\
0 & 0 & e^{-(2\gamma_0-i\Omega)\Delta t} & e^{-(2\gamma_0-i\Omega)\Delta t}
\end{pmatrix},
$$

$$
\tag{7.121}
$$

$$
e^{\mathscr{L}_A \Delta t} =
\begin{pmatrix}
1 & 0 & 0 & 0 \\
0 & 1 & 0 & 0 \\
0 & 0 & e^{-(2\gamma_0+i\Omega)\Delta t} & 0 \\
0 & 0 & 0 & e^{-(2\gamma_0-i\Omega)\Delta t}
\end{pmatrix}.
$$

7.5.3 Phenomenological Consequences

Even without any decoherence ($\gamma_0 = 0$), this may have significant consequences: the repeated application of the propagator for measurement without feedback ($\gamma_0 = 0$ and $\alpha = 0$) yields

$$
\left[e^{\mathscr{L}_A \Delta t}(\mathscr{P}_1 + \mathscr{P}_2) \right]^n
$$

$$
= \frac{1}{2}
\begin{pmatrix}
1 & 1 & 0 & 0 \\
1 & 1 & 0 & 0 \\
0 & 0 & e^{-i\Omega \Delta t}\cos^{n-1}(\Omega \Delta t) & e^{-i\Omega \Delta t}\cos^{n-1}(\Omega \Delta t) \\
0 & 0 & e^{+i\Omega \Delta t}\cos^{n-1}(\Omega \Delta t) & e^{+i\Omega \Delta t}\cos^{n-1}(\Omega \Delta t)
\end{pmatrix}.
\tag{7.122}
$$

Exercise 7.9 (Repeated measurements) Show the validity of the above equation.

In contrast, without the measurements, we have for repeated application of the propagator simply

$$
\left[e^{\mathscr{L}_A \Delta t} \right]^n = e^{\mathscr{L}_A n \Delta t}.
\tag{7.123}
$$

When we now consider the limit $n \to \infty$ and $\Delta t \to 0$ but $n\Delta t = t$ remaining finite, it becomes obvious that the no-measurement propagator for $\gamma_0 = 0$ simply describes coherent evolution. In contrast, when the measurement frequency becomes large enough, the measurement propagator in Eq. (7.121) approaches

$$
\left[e^{\mathscr{L}_A \Delta t}(\mathscr{P}_1 + \mathscr{P}_2) \right]^n = \frac{1}{2}
\begin{pmatrix}
1 & 1 & 0 & 0 \\
1 & 1 & 0 & 0 \\
0 & 0 & 1 & 1 \\
0 & 0 & 1 & 1
\end{pmatrix}
\tag{7.124}
$$

and thereby freezes eigenstates of the measurement operators such as $\bar{\rho} = \frac{1}{2}[|0\rangle + |1\rangle][\langle 0| + \langle 1|]$. This effect is known as the quantum Zeno effect [20, 21] (sometimes colloquially phrased as "a watched pot never boils") and occurs when measurement operators and system Hamiltonian do not commute and the evolution between measurements is unitary (here $\gamma_0 = 0$).

Unfortunately, when the evolution between measurements is an open one ($\gamma_0 > 0$), the quantum Zeno effect cannot be used to stabilize the coherences, which becomes evident from the propagator in Eq. (7.121). Although the effective decoherence rate will be reduced, in the long-term dynamics only the populations will survive.

With feedback ($\alpha > 0$), the effective propagator $\mathcal{W}(\Delta t)$ does not have the block structure of the Liouvillians anymore. The propagator defines a fixed-point iteration for the density matrix,

$$\rho(t + \Delta t) = \mathcal{W}(\Delta t)\rho(t). \tag{7.125}$$

Here, even for small Δt we cannot approximate the evolution by another effective Liouvillian, since $\lim_{\Delta t \to 0} \mathcal{W}(\Delta t) \neq \mathbf{1}$; i.e., a master equation description is not applicable. Instead, one could in principle analyze the eigenvector of $\mathcal{W}(\Delta t)$ with eigenvalue 1 as the (in a stroboscopic sense) stationary state. For the present model, it is however more convenient to consider the expectation values of $\langle \sigma^i \rangle_t$ that fully characterize the density matrix via

$$\rho_{00} = \frac{1 + \langle \sigma^z \rangle}{2}, \qquad \rho_{11} = \frac{1 - \langle \sigma^z \rangle}{2},$$
$$\rho_{01} = \frac{\langle \sigma^x \rangle - i\langle \sigma^y \rangle}{2}, \qquad \rho_{10} = \frac{\langle \sigma^x \rangle + i\langle \sigma^y \rangle}{2}, \tag{7.126}$$

which is known as Bloch sphere representation. Note that decoherence therefore implies vanishing expectation values of $\langle \sigma^x \rangle \to 0$ and $\langle \sigma^y \rangle \to 0$ in our setup. Converting the iteration equation for the density matrix into an iteration equation for the expectation values of Pauli matrices, we obtain

$$\langle \sigma^x \rangle_{t+\Delta t} = \frac{e^{-2\gamma_0 \Delta t}}{2} \Big\{ \big(1 + \langle \sigma^x \rangle_t\big) \cos(\Omega \Delta t)$$
$$- \big(1 - \langle \sigma^x \rangle_t\big) e^{-(\gamma_- + \gamma_+)\alpha \Delta t/2} \cos\big[(\Omega + \alpha(\Omega + \Sigma))\Delta t\big] \Big\},$$

$$\langle \sigma^y \rangle_{t+\Delta t} = \frac{e^{-2\gamma_0 \Delta t}}{2} \Big\{ \big(1 + \langle \sigma^x \rangle_t\big) \sin(\Omega \Delta t) \tag{7.127}$$
$$- \big(1 - \langle \sigma^x \rangle_t\big) e^{-(\gamma_- + \gamma_+)\alpha \Delta t/2} \sin\big[(\Omega + \alpha(\Omega + \Sigma))\Delta t\big] \Big\},$$

$$\langle \sigma^z \rangle_{t+\Delta t} = \frac{(\gamma_+ - \gamma_-)(1 - \langle \sigma^x \rangle_t)}{2(\gamma_- + \gamma_+)} \big(1 - e^{-(\gamma_- + \gamma_+)\alpha \Delta t}\big),$$

where we note that—since we have chosen to measure the value of σ^x—both $\langle \sigma^y \rangle_t$ and $\langle \sigma^z \rangle_t$ follow just the expectation value of $\langle \sigma^x \rangle_t$. The first of the above equations

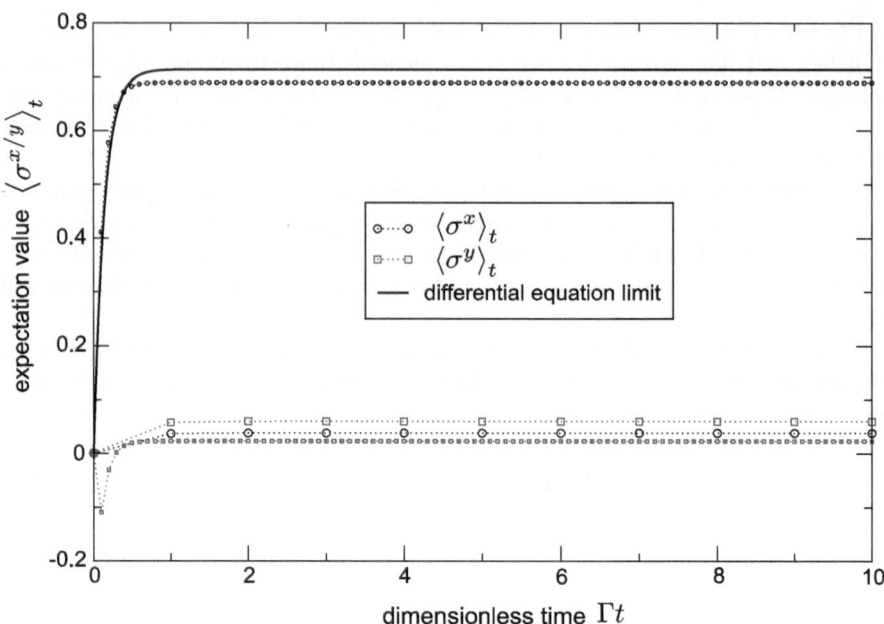

Fig. 7.15 Expectation values of the Pauli matrices for finite feedback strength $\alpha = 10$ and finite stepsize Δt (spacing given by *symbols*). For large Δt, the *fixed point* corresponds to a nearly completely mixed state close to the center of the Bloch sphere. For small Δt, the curve for $\langle \sigma^x \rangle_t$ approaches the differential equation limit (*solid line*), but the curve for $\langle \sigma^y \rangle_t$ approaches 0. For $\gamma_- = \gamma_+$, the iteration for $\langle \sigma^z \rangle_t$ vanishes throughout. Parameters: $\gamma_- = \gamma_+ = \gamma_0 = \Gamma$, $\Omega \Delta t = \in \{1, 0.1\}$, and $\Sigma \Delta t \in \{0.5, 0.05\}$

can be expanded for small Δt to yield

$$\frac{\langle \sigma^x \rangle_{t+\Delta t} - \langle \sigma^x \rangle_t}{\Delta t} = -\frac{1}{4}\left[8\gamma_0 + \alpha(\gamma_- + \gamma_+)\right]\langle \sigma^x \rangle_t$$

$$+ \frac{1}{4}\alpha(\gamma_- + \gamma_+) + \mathcal{O}\{\Delta t\}. \tag{7.128}$$

When $\Delta t \to 0$, this becomes a differential equation:

$$\frac{\langle \sigma^x \rangle_{t+\Delta t} - \langle \sigma^x \rangle_t}{\Delta t} \to \langle \dot{\sigma}^x \rangle_t \tag{7.129}$$

with the stationary state

$$\langle \bar{\sigma}^x \rangle = \frac{\alpha(\gamma_- + \gamma_+)}{8\gamma_0 + \alpha(\gamma_- + \gamma_+)}, \tag{7.130}$$

which approaches 1 for large values of α, i.e., for sufficiently strong feedback. In the same limit, the other expectation values simply vanish throughout $\langle \sigma^{y/z} \rangle_t \to 0$, which is just a consequence of the large measurement rate in the x-direction. Taking into account the large-temperature expansions for the damping coefficients,

$$\gamma_0 = J_0 k_B T, \qquad \gamma_- + \gamma_+ \approx 2 J_0 e^{-\Omega/\omega_c} k_B T, \qquad (7.131)$$

we see that this stabilization effect also holds at large temperatures—with a sufficiently strong (and perfect) feedback provided. An initially coherent superposition is thus not only stabilized, but also emerges as a stroboscopic stationary state when the scheme is initialized in a completely mixed state, e.g., at the center of the Bloch sphere $\langle \sigma^\alpha \rangle = 0$. Also for finite Δt, the fixed-point iteration yields sensible evolution for the expectation values of the Pauli matrices; see Fig. 7.15. So formally, we have a similar situation as in Sect. 7.3—also known as Parrondo's paradox [8]: two losing strategies (each Liouvillian applied alone leads to decoherence) can be combined, together yielding (with the measurement) a winning strategy (avoiding decoherence). The fact that decoherence can be useful in these setups can also be understood in terms of the simple interpretation that undesired parts of the density matrix (those that are not eigenstates of the measurement operators) are damped away faster than the eigenstates if the measurements occur frequently enough [22].

References

1. T. Brandes, Feedback control of quantum transport. Phys. Rev. Lett. **105**, 060602 (2010)
2. L. Fricke, F. Hohls, N. Ubbelohde, B. Kaestner, V. Kashcheyevs, C. Leicht, P. Mirovsky, K. Pierz, H.W. Schumacher, R.J. Haug, Quantized current source with mesoscopic feedback. Phys. Rev. B **83**, 193306 (2011)
3. H. Leff, A. Rex, *Maxwell's Demon: Entropy, Information, Computing* (Hilger, Bristol, 1990)
4. R. Landauer, Irreversibility and heat generation in the computing process. IBM J. Res. Dev. **5**, 183 (1961)
5. D.V. Averin, M. Möttönen, J.P. Pekola, Maxwell's demon based on a single-electron pump. Phys. Rev. B **84**, 245448 (2011)
6. A. Berut, A. Arakelyan, A. Petrosyan, S. Ciliberto, R. Dillenschneider, E. Lutz, Experimental verification of Landauer's principle linking information and thermodynamics. Nature **483**, 187 (2012)
7. O.-P. Saira, Y. Yoon, T. Tanttu, M. Möttönen, D.V. Averin, J.P. Pekola, Test of the Jarzynski and crooks fluctuation relations in an electronic system. Phys. Rev. Lett. **109**, 180601 (2012)
8. G.P. Harmer, D. Abbott, Losing strategies can win by Parrondo's paradox. Nature **402**, 864 (1999)
9. G. Schaller, C. Emary, G. Kiesslich, T. Brandes, Probing the power of an electronic Maxwell's demon: single-electron transistor monitored by a quantum point contact. Phys. Rev. B **84**, 085418 (2011)
10. T. Sagawa, M. Ueda, Second law of thermodynamics with discrete quantum feedback control. Phys. Rev. Lett. **100**, 080403 (2008)
11. T. Sagawa, M. Ueda, Minimal energy cost for thermodynamic information processing: measurement and information erasure. Phys. Rev. Lett. **102**, 250602 (2009)
12. T. Sagawa, M. Ueda, Generalized Jarzynski equality under nonequilibrium feedback control. Phys. Rev. Lett. **104**, 090602 (2010)
13. T. Sagawa, M. Ueda, Nonequilibrium thermodynamics of feedback control. Phys. Rev. E **85**, 021104 (2012)
14. C. Emary, Delayed feedback control in quantum transport. Philos. Trans. R. Soc. A **371**, 1471 (2013)

15. P. Strasberg, G. Schaller, T. Brandes, M. Esposito, Thermodynamics of quantum-jump-conditioned feedback control. Phys. Rev. E **88**, 062107 (2013)
16. P. Strasberg, G. Schaller, T. Brandes, M. Esposito, Thermodynamics of a physical model implementing a maxwell demon. Phys. Rev. Lett. **110**, 040601 (2013)
17. U. Seifert, Entropy production along a stochastic trajectory and an integral fluctuation theorem. Phys. Rev. Lett. **95**, 040602 (2005)
18. R. Sánchez, M. Buttiker, Optimal energy quanta to current conversion. Phys. Rev. B **83**, 085428 (2011)
19. D.P. DiVincenzo, The physical implementation of quantum computation. Fortschr. Phys. **48**, 771 (2000)
20. B. Misra, E.C.G. Sudarshan, The Zeno's paradox in quantum theory. J. Math. Phys. **18**, 756 (1977)
21. P. Facchi, S. Pascazio, Quantum Zeno subspaces. Phys. Rev. Lett. **89**, 080401 (2002)
22. G. Schaller, Fighting decoherence by feedback-controlled dissipation. Phys. Rev. A **85**, 062118 (2012)

Index